QUANTITATIVE EVALUATION OF SPECTROGRAMS BY MEANS OF *l*-TRANSFORMATION

QUANTITATIVE EVALUATION OF SPECTROGRAMS

BY MEANS OF *l*-TRANSFORMATION

by

TIBOR TÖRÖK D. Sc.
Professor of Chemistry

and

KÁROLY ZIMMER D. Sc.
Senior Lecturer

Institute for Inorganic and Analytical Chemistry
of the L. Eötvös University, Budapest

HEYDEN & SON LTD

LONDON · NEW YORK · RHEINE

Translated by

L. NEMES C. Sc.

Heyden & Son Ltd., Spectrum House, Alderton Crescent, London NW4 3XX
Heyden & Son Inc., 225 Park Avenue, New York, N.Y. 10017, U.S.A.
Heyden & Son GmbH, Steinfurter Strasse 45, 4440 Rheine/Westf., Germany

ISBN 0 85501 062 2

Joint edition with
Akadémiai Kiadó, Budapest

Printed in Hungary

CONTENTS

[redacted]

FOREWORD

The great variety, complex composition or even the required high purity of the materials used in modern industry are characteristic concomitants of the scientific and technical revolution of our age. At the same time requirements in the mechanical, physical and chemical properties of structural materials have also increased. Demands on test methods have grown correspondingly. Analytical methods of high performance, sensitivity and accuracy have become necessary.

In keeping pace with the demands of industry, the development of science has provided new methods for satisfying the different needs, and, on the other hand, has extended the applicability of procedures which have already been in standard use. The new methods have developed in many ways: they improve the limit of detection, increase the accuracy, dependability, speed, simplicity, the scope and economy of the determinations.

Among the great number of physical and chemical methods for the determination of a wide range of elements, chemical emission spectral analysis is very useful and appropriate. The classic *spectroscopical method* is still widely used in the less demanding metallurgical examinations. The low price, easy transportability and convenient handling of the equipment are of great advantage. The procedure is rapid, and it is applicable, first of all, for recognizing accidental mixing of materials.

The branch of spectral analysis that has undergone most spectacular development in the last decade is *spectrometry.* No other spectrochemical procedure can provide the accuracy and speed of spectrometry. In all sorts of metallurgical laboratories and, recently, in the study of geological samples and also in other fields this method can be used successfully for the routine analysis of a large number of samples of known type. Non-metallic elements can also be conveniently determined by means of *vacuum spectrometers.* The relatively complicated calculations are done quickly and reliably by computers connected to the apparatus.

The conventional method of spectral analysis, the *spectrographical procedure,* has not lost its significance, either. It is the only useful spectrochemical method for the study of materials of unknown composition. Its accuracy is much higher than that of visual spectroscopical examinations, and its limit of detection, in general, surpasses that of spectrometric analysis. To acquire, to set up and to operate these instruments is much cheaper than in the case of spectrometers, and there are a great number of laboratories where everyday analyses can be done by means of a spectrograph. As opposed to spectrometric analysis, the spectrograms also show the lines of the elements whose presence has not even been expected in the sample to be analyzed. The spectrum plate itself makes it possible to check the analyses at a later date, too, which is important in disputable cases. The spectrograph is a necessary instrument in spectrometric laboratories, too. Its uses, among other things, are calibration, checking, special analyses and development studies. Spectrographic studies can be very easily suited to the most different requirements; the type of excitation, the quality of the emulsion can be readily altered to fit the particular purpose. Using a medium dispersion instrument, the full analytically important optical spectral range can be photographed, and with high-dispersion spectrographs the necessary wavelength region can be selected easily.

Spectrographic methods find steadily increasing applications. There is a great number of papers in the literature, research results of theoretical nature, which deal with the different developments and practical applications of this method.

The purpose of our work is to help those working in the field of quantitative spectral analysis in spectrographical laboratories. Our book contains the numerical tables needed for the practical calculations of l-transformation. In the preceding text the most important methods of blackening transformation are described, with special regard to the development of l-transformation. Only the final results are summarized for methods found scattered throughout the literature, but it was attempted to give a possibly complete list of the relevant literature. Each step of the practical performance of l-transformation is given in detail. The full procedure of quantitative spectrographical evaluation is illustrated by means of practical examples.

We hope that the present book will be helpful to those working in the field, resulting in greater simplicity, rapidity, and reliability of their work.

Budapest, January, 1970

The Authors

1. THEORETICAL

1.1 INTRODUCTION

The relative accuracy of spectrographic analyses is constant within wide limits of concentrations. Consequently, *the absolute analytical error is smaller at lower concentrations of the elements to be determined.*

Because of the above reasons and owing to practical requirements, in recent years spectrographical analysis has acquired increasing importance mainly in *trace analysis*. In the case of low concentrations very frequently low blackening values are measured, and the composition of the sample is deduced therefrom. This, however, gives rise to problems in evaluation, since low blackening values correspond to the lower, curved part of the characteristic curve. On this part—as opposed to the straight part of the characteristic curve—there is no linear relationship between the blackening, S, and the logarithm of intensity, $\lg I$ (see Fig. 1.4.1. p. 5). Therefore, there is no direct proportionality between the line blackening and the logarithm of concentration, which renders the evaluation difficult.

Accordingly, the measured blackening values cannot be *directly* applied for analytical purposes without the use of etalons. It is necessary for each spectral plate to photograph the spectra of standards, which are very expensive and generally difficult to acquire. This also causes a significant loss of time in completing the analyses.

In order to simplify evaluation, it is necessary to find a definite functional relationship between the blackening measured and another quantity which is linearly related to the logarithm of intensity. This quantity is called *transformed blackening*. The operation itself by means of which the blackening values are recalculated to transformed blackening values is called *blackening transformation*. Modern quantitative spectrographic evaluations are inconceivable today without the use of blackening transformation methods.

If the following a short review of the development of blackening transformation methods is given, then l-transformation will be described in detail. The underlying principles of the equations are only discussed to an extent necessary for an understanding of the relationships. No theoretical treatment of the processes taking place in the photographic emulsion will be given.

1.2 LIST OF SYMBOLS [1]

a	factor of the Scheibe–Lomakin equation
a	*in subscript*: refers to the unweakened step of spectra produced by two-step filter or sector
α	angle between the straight part of the characteristic curve and the abscissa
b	exponent of the Scheibe–Lomakin equation, i.e. the slope of the analytical curve of the function ΔY, lg c
b	*in subscript*: refers to the weakened step of spectra produced by two-step filter or sector
β	exponent, characteristic of the absorption, in the Scheibe–Lomakin–Kerekes equation
c	concentration
c'	$= c_x/c_r$, relative concentration
c_i	concentration of foreign element
c'_m	relative concentration corresponding to the ΔY_m value
c'_r	the Gerlach homologous concentration
d	$= s - \lg(10^s - 1)$, Gaussian subtraction logarithm belonging to s
D	$= S - \lg(10^S - 1)$, Gaussian subtraction logarithm belonging to S
Δ	difference between two values
η	slope of the analytical curve of the form: lg c, ΔY
γ	gradation, the slope of the straight portion of the characteristic curve
I	radiation intensity
k	constant of l-transformation
$\varkappa$	constant of $P_\varkappa$-transformation
$\varkappa'$	constant of P_L-transformation
$\varkappa_0$	one of the constants of L-transformation
L	L-transformed blackening
l	l-transformed blackening
Λ	Λ-transformed blackening
m	one of the constants of the modified Kaiser approximation
n	any integral number
ω	central angle of the sector opening of the step sector
P	general symbol of transformed blackening, one of the usual designations for $P_\varkappa$-transformation
P_a	a former symbol for $P_\varkappa$ transformation
$P_\varkappa$	$P_\varkappa$-transformed blackening
P'_x	modified $P_\varkappa$-transformed blackening
P_L	P_L-transformed blackening
P_l	one of the former symbols for $P_\varkappa$-transformation
P_W	the generalized designation for W-transformation
$P_{1/2}$	$P_{1/2}$-transformed blackening
q	sector- or filter constant, respectively
r	*in subscript*: reference (standard)
S	blackening
S_L	lower limit for the straight portion of the characteristic curve

ΔS_L	length of the straight portion of the characteristic curve
ΔS_m	blackening difference belonging to the pair of intensity-calibration pattern
s	$= S/\gamma$ reduced blackening
T	transmittance (transparency)
T_p	photographic transmittance
u	*in subscript*: background
$u(r)$	background beside the line of the reference element
$u(x)$	background beside the analytical line
W	the usual designation of W-transformation
x	*in subscript*: unknown
Y	$= \lg I$ logarithm of intensity
ΔY_m	the logarithmic filter- or sector constant, i.e. the negative logarithm of the filter- or sector constant
$'$	*in superscript*: refers to the spectrum of higher intensity
$''$	*in superscript*: refers to the spectrum of lower intensity

1.3 RELATIONSHIP BETWEEN CONCENTRATION AND THE INTENSITY RATIO OF THE ANALYTICAL LINE-PAIR

Photographic emulsions exposed to the effect of rays of appropriate intensity and wavelength are blackened on development. *The extent of blackening of the spectral lines*—as measured on a spectral plate or film—*depends on the intensity of the illuminating radiation.*

There is a definite relationship between the concentration, c, of element x present in the sample and the intensity, I, of its analytical line, under identical experimental conditions. This relationship is described by the fundamental equation of quantitative spectral analysis, the *Lomakin–Scheibe* empirical equation [2–4]:

$$I = a \cdot c^b$$

where a and b are parameters depending on the experimental conditions.

In practice the intensity I_x of the analytical line of element x to be determined is related to the intensity I_r of the radiation arising from the ground element (matrix) or from some other *r*eference element or material r, in order to reduce the error due to variations in the experimental conditions. The following formula is obtained by writing the above equation for the intensity relationships:

$$\frac{I_x}{I_r} = a' \left(\frac{c_x}{c_r}\right)^b$$

where $a' = a_x/a_r$.

This equation is usually applied in logarithmic form in practical spectral analysis. If Y is used, as suggested by Kaiser [5], to denote the logarithm of intensity and ΔY for their differences, we obtain:

$$\Delta Y = \lg \frac{I_x}{I_r} = \lg a' + b \cdot \lg \frac{c_x}{c_r}\,.$$

Considering that the value measured in practice is ΔY and the concentration is derived from this, the above equation is most useful written in the following form:

$$\lg c' = \lg c_r' + \eta \Delta Y \tag{1.3.1}$$

where $c' = c_x/c_r$ is the relative concentration;

$$c_r' = \left(\frac{1}{a'}\right)^{\frac{1}{b}}$$

is the Gerlach homologous concentration [6], at which the intensities of the two members of the analytical line-pair are identical, i.e. $\Delta Y = 0$; finally, $\eta = 1/b$ is the reciprocal slope value of the analytical curve $\lg c'$ vs. ΔY. Thus experience also shows that a linear relationship exists between the logarithm of the radiation intensity ratio and the logarithm of the relative concentration. This is indicated by the fact that the analytical curve constructed on the basis of Eq. (1.3.1) is a straight line between not too wide concentration limits (Fig. 1.3.1).

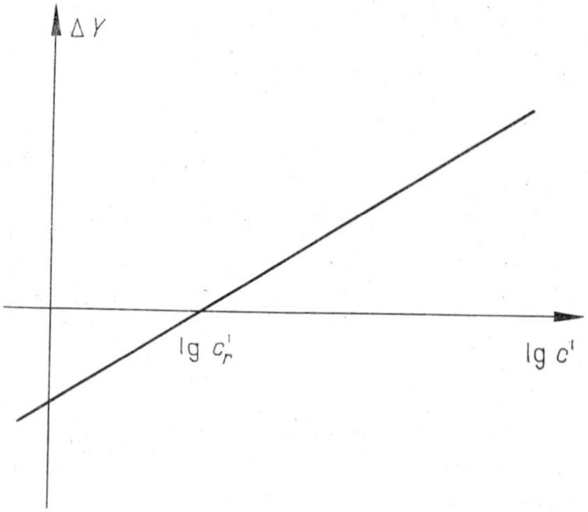

FIG. 1.3.1. The analytical curve

At high concentrations, however, the analytical curve is no longer linear, but bent in the direction of the concentration axis. The reason for this is that *absorption occurs in addition to emission.* The hot, radiating plasma is enclosed within a non-radiating cold envelope. The rays emitted by the plasma have first to pass through this envelope, where the atoms are found in ground state, corresponding to the lower temperature. Here absorption occurs inevitably, its extent being proportional to the concentration of the absorbing atoms.

The effect of absorption is taken into account by the theoretically well founded [7] Scheibe–Lomakin–Kerekes equation [8]:

$$I = a \cdot c^b \cdot c^{-\beta c}$$

where β is a parameter independent of the concentration and characteristic of the absorption.

Self-absorption can be avoided in practice by the suitable choice of the analytical line, according to the concentration range in question. The lines specified in the analytical prescriptions normally meet the above condition

1.4 RELATIONSHIP BETWEEN THE INTENSITY AND THE BLACKENING OF THE SPECTRAL LINES

The complicated relationship between the intensity of the emitted radiation and the blackening of the emulsion brought about by this radiation is not yet known exactly. As suggested first by Hurter and Driffield [9],

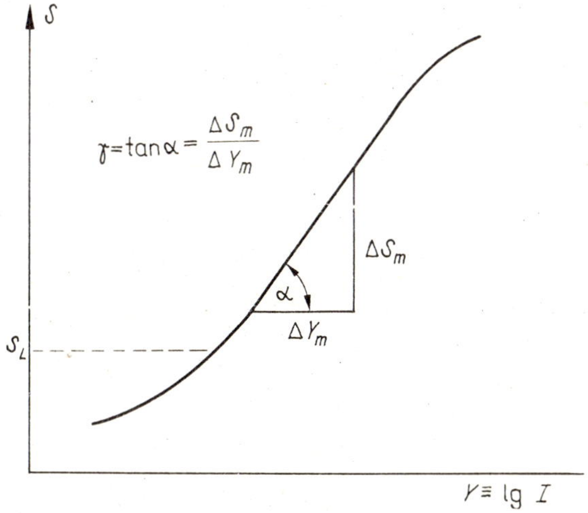

FIG. 1.4.1. The blackening curve

the density is plotted as a function of the logarithm of exposure, or when the exposure time and the cross-section of the beam are constant (in spectral analysis this is generally true), the plot is made using the logarithm of the light intensity. This plot is called the *characteristic curve* (Fig. 1.4.1). This way of plotting is advantageous because *in the region of medium blackening values the relationship between blackening and the logarithm of intensity is linear.*

The concentration can be simply determined from the blackening values corresponding to the straight part of the characteristic curve, because the blackening difference of the analytical line-pair is proportional to the logarithm of the corresponding line intensity ratio:

$$\Delta S = \gamma \cdot \lg \frac{I_x}{I_r} = \gamma \cdot \Delta Y. \qquad (1.4.1)$$

In this equation γ is the slope of the straight part of the characteristic curve.

When the above formula is substituted into Eq. (1.3.1) the following relationship is obtained:

$$\lg c' = \lg c'_r + \frac{\eta}{\gamma} \Delta S. \tag{1.4.2}$$

It is obvious from the above equation that *there is a linear relationship between the logarithm of the element concentration and the blackening differences measured on the straight part of the characteristic curve.*

In such cases the evaluation can be done quickly, and there is no need to photograph comparison spectra onto each spectrum plate, provided the values of quantities γ, η and c'_r have been determined previously, i.e. when they are known. Thereby time, labour and material are saved.

The growing importance of trace analysis has, however, made necessary the determination of lower blackening values. Improvements in the excitation and optical conditions cannot increase the radiation intensity on the spectrum plate beyond a certain limit. The application of high-sensitive emulsions is limited, among other factors, by the fact that when the sensitivity is increased, the scatter of blackening values also increases. *When the element concentration is low, the blackening on the emulsion will be slight, and these values correspond to the lower, curved part of the characteristic curve.* In other words, *the low concentration part of the evaluation curve will become curved.*

1.5 BLACKENING TRANSFORMATION METHODS

In order to eliminate the above difficulties, it is necessary to find a sufficiently accurate method of the spectrographic determination of low concentrations without the incessant use of standards. This task has been solved by the introduction of blackening transformation methods.*

The essential feature of blackening transformation is that it assigns such a quantity to the transmittance (T) or to the blackening (S), as—analogously to Eqs. (1.4.1) and (1.4.2)—*remains linearly proportional to the logarithm of intensity (Y), thereby to the logarithm of concentration as well,* even when the blackening value is not on the straight part of the characteristic curve.

* Blackening, S, is a quantity derived from the primary concept of photographic transmittance T_p (transparency):

$$S = -\lg T_p.$$

In European spectrochemical laboratories almost exclusively blackening is measured, while in the United States mainly transmittance is used. Therefore, the most important relations will be given in the following on the basis of transmittance as well.

On the scale of densitometers the hundredfold values of the actual blackening and transmittance can be read.

The term 'blackening' frequently occurs in the English literature in the form of 'density'. The use of this term is, however, justified only when it is preceded by the specification 'optical' or 'photographic'. To avoid misunderstanding, only the term 'blackening' is used in the present book, according to recent recommendations (see, footnote on p. 8). A further reason for doing so is that in connection with transformation procedures the expression: 'blackening transformation' is widely applied.

Blackening transformation can be effected both graphically and by numerical calculation. The graphical solution is very simple. First the characteristic curve [10, 11] is constructed using the blackening values measured on the wavelength of the analytical line under the given experimental conditions (excitation, optical, photographical and evaluation conditions). The straight part of the characteristic curve is elongated towards the low intensity values. This straight line which also includes the straight part of

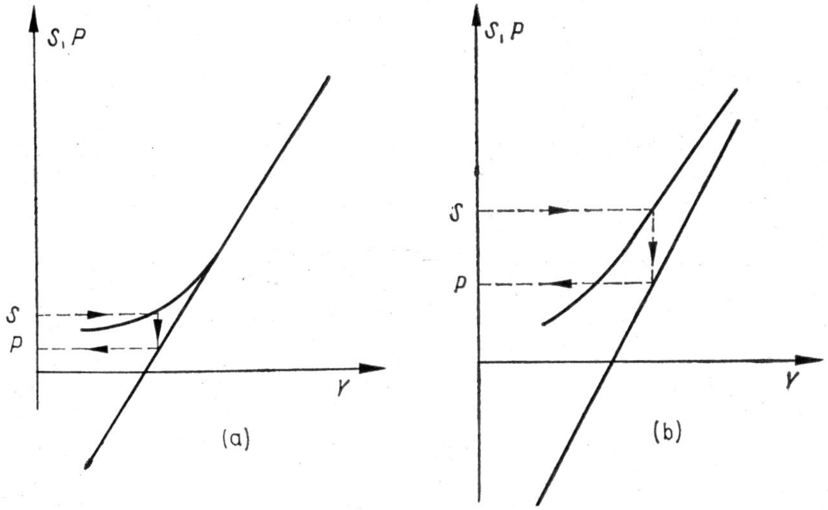

FIG. 1.5.1. Two theoretical possibilities of the graphical blackening transformation

the characteristic curve, is the *transformed* or 'straightened' *characteristic curve* (Fig. 1.5.1(a)). In principle any straight line, obtained either by choice or by calculation, can be used as transformed characteristic curve (Fig. 1.5.1(b)).

The transformed blackening value belonging to a blackening S is denoted by P. The related S, P values are given by the corresponding values belonging to the points on the same ordinate of the characteristic curve. The transformation is done by projecting the measured blackening values from the characteristic curve to the previously drawn straight line, parallel to the ordinate, thereafter reprojecting parallel to the abscissa (see Fig. 1.5.1). The value read at this point is the transformed blackening [12]. In spite of its simplicity, this method has not gained general acceptance, since the construction of individual characteristic curves would be difficult and time-consuming in routine analyses.

1.5.2 SAMPSON–BAKER–SEIDEL TRANSFORMATION

Blackening transformation by calculation is of earlier origin. Sampson [13] and Baker [14, 15] and also Seidel [16] observed several decades ago that the lower curved part of the characteristic curve can be 'straightened' if a correction term is deducted from the blackening values; this term is a definite function of blackening:

$$P_W = S - D$$

where P_W is the transformed blackening,* the correction denoted by D is the Gaussian subtraction logarithm, belonging to S, whose value is given by:

$$D = S - \lg (10^S - 1) .$$

This relationship has an important role in the different fields of spectrographic evaluation (P- and l-transformation, background correction).

Combination of the above two equations gives the relationship between the quantities P_W and S:

$$P_W = \lg (10^S - 1).$$

The relationship between P_W and T_p is:

$$P_W = \lg \left(\frac{1}{T_p} - 1\right) .$$

The W-transformation was the first method which—on the ground of purely empirical observations—assigned such a quantity to the measured blackening which is in linear relation with the Y value. According to more detailed investigations [17] this method can only be applied for certain types of emulsions, and even then, only within a very narrow range of wavelengths (e.g. in the case of Agfa blau extrahart emulsions, at about 370 nm).

1.5.3 HONERJÄGER-SOHM–KAISER TRANSFORMATION

Starting with the observation that the W-transformation overcorrects in the medium ultraviolet range, which is very important from the point of view of the analytical lines of technically important metals, Honerjäger-Sohm and Kaiser [17] introduced the simple modification that only half of the D value was deduced from the measured blackening values. Accord-

* The original and usual designation of the Sampson–Baker–Seidel transformation is W. The IUPAC Committee in its proposition 'Nomenclature, Symbols and Usage in Analytical Atomic Spectroscopy, Part I', published in 1967, suggested the symbol P (*p*hotographic *p*arameter) for the designation of transformed blackening with an index denoting the transformation method applied. In order to give a consistent treatment, this designation is used in this book too, when procedures similar to the original P-transformation are used.

ingly, the $P_{1/2}$-transformation suggested by the latter authors (the index was introduced later) is given by the following equation:

$$P_{1/2} = S - \frac{1}{2} D.$$

This procedure can be applied in a wider spectral range — which is just the one important in spectrographic analysis — between 250 and 320 nm, and generally for a number of emulsion types. However, this method cannot be applied for analytical lines of other wavelengths.

1.5.4 KAISER TRANSFORMATION

The multiplication factor of the D correction is 1 for the W-transformation, while it is 1/2 for the $P_{1/2}$-transformation. When, instead of the above values, a transformation coefficient is introduced whose value depends on the type of the photographic plate, method of development and wavelength used [18], the blackening transformation will hold good for wider limits [17]. The general P-transformation [19] — whose designation was originally P_a, P_l and finally $P_\varkappa$ — has the following final form [20] written for blackening (S) and photographic transmittance [T_p], respectively:

$$P_\varkappa = S - \varkappa \cdot D \tag{1.5.1}$$

and

$$P_\varkappa = \varkappa \lg (1 - T_p) - \lg T_p.$$

The $P_\varkappa$-transformation has found wide application since it gives, in general, good accuracy and is relatively simple. In many cases the exact determination of $\varkappa$ has also given rise to problems. This is generally done so [19] that the blackening of the corresponding lines of the spectra, obtained with a two-step filter (or two-step rotating sector, etc.), is measured at two steps of the spectra. Equation (1.5.1) gives the difference of these blackening values:

$$\Delta P\varkappa = \Delta S - \varkappa \Delta D. \tag{1.5.2}$$

If the blackening values of a pair $\Delta S = S_a - S_b$ are situated on the straight part of the characteristic curve, while those of another value-pair ΔS (belonging to the same intensity signal-pair) lie on the under-exposed part [21], $\varkappa$ can be determined from the equations, since $\Delta P_\varkappa$ is identical in both cases. On the basis of the following formula [19]:

$$\varkappa = \frac{\Delta S' - \Delta S''}{\Delta D' - \Delta D''} = \frac{\Delta \Delta S}{\Delta \Delta D}$$

$\varkappa$ can only be determined, however, with low precision due to the double differences. This can be improved by averaging a great number of data [11].

The method proposed by Boumans [22] has also proved advantageous. Equation (1.5.2) is the equation of a straight line. When the ΔD-values are plotted as function of the blackening differences ΔS and a straight line is fitted to the experimental points, the slope of this straight line gives directly $\varkappa$.

The $P_\varkappa$-transformation is an empirical equation. The transformation coefficient is not associated with a physical meaning. This is the actual weight factor for the correction term D. More detailed studies [21] have shown that the $P_\varkappa$-transformed blackening curve is not strictly a straight line. At the point of the characteristic curve where the under-exposed and the straight parts meet, there is an inflection or break point on the transformed curve, which divides the transformed characteristic curve into two sections of somewhat different slopes [23]. This obviously involves an increase of error. $P_\varkappa$-transformation renders the straight part of the original characteristic curve slightly curved, since non-linearly varying $\varkappa D$-values will be deducted from the S-values which vary linearly as a function of Y. For blackening values smaller than 0.10–0.05 the $P_\varkappa$-transformation deviates quickly from the straight line. Due to these drawbacks, the method cannot be regarded as a final solution of the problems of blackening transformation, in spite of all of its advantages.

1.5.5 MODIFIED KAISER TRANSFORMATION

The transformation factor $\varkappa$ depends on the wavelength [18, 19] and even on the absolute value of the blackening [24–27]. Kaiser [19] took this blackening dependence into account by introducing a new correction factor.

The equation of Kaiser's second approximation is:

$$P'_\varkappa = S - \varkappa D - m\frac{D}{S^2}$$

where m is a factor depending on the wavelength. Since the value of m should be determined by trial and error, the equation containing the two correction factors is inconvenient and has not found application in practice. Its accuracy is, however, an advantage, when an appropriately programmed computer is used for the calculations.

1.5.6 CANDLER TRANSFORMATION

Candler [28] sought an equation of linear transformation which is independent of the wavelength. The Λ-transformation which was introduced by him can be regarded as a $P_\varkappa$-transformation divided by $\varkappa$ [18]. Accordingly, the accuracy of the Λ-transformation is determined by the scatter in the determination of $\varkappa$, which—as we have seen—is rather large. The condition of wavelength-independence is only fulfilled if the ratio of $\varkappa$ and γ is a constant throughout the full spectral range [18]. For certain emulsion types (e.g., Agfa and Orwo blau extrahart) the ratio $\varkappa/\gamma$ is a constant, while for other spectrum plates (e.g., Perutz and Ilford) it is not [29, 30].

1.5.7 FURTHER RESEARCH

Further research has been directed partly toward the development and perfection of the $P_\varkappa$-transformation, and towards deriving new transformation equations. Crosswhite and Dieke [31], furthermore Arrak [32] proposed transformation functions which are related to the $P_\varkappa$-transformation.

Most transformations do not utilize the advantage provided by the original characteristic curve, which is that the straight part can be directly used for the evaluation. These procedures also transform the straight part of the characteristic curve, in other words, they 'spoil' it, thereby giving rise to further error.

1.5.8 MORELLO TRANSFORMATION

The aim of further investigations was, therefore, to construct a transformation equation only for the curved part of the characteristic curve. Morello [33, 34] has described the lower part of the characteristic curve by a second-order equation. It is a great advantage of Morello's empirical relationship that this transformation gives good results if the values of the constants are chosen correctly. It is, on the other hand, a problem that the equation contains four parameters: S_L, the lower limit of the straight part of the characteristic curve; γ, and two factors depending on the properties of the emulsion. The determination of these parameters is laborious, thus the application of the equation is not very convenient.

1.5.9 P_L-TRANSFORMATION

The P_L-transformation (originally, L-transformation) is applied to achieve an elongation of the straight part of the characteristic curve towards the low blackening values [35]. The straight part of the characteristic curve and the corresponding part of the transformed characteristic curve are identical. The fundamental equation (Eq. 1.5.3) of the $P_\varkappa$-transformation remains formally the same. Accordingly, the value of $\varkappa$ deviates from the original definition [36]. The P_L-transformation is, however, not simple enough, and its application is time-consuming.

1.5.10. L-TRANSFORMATION

The following modification of the P_L-transformation has led to surprisingly good results [37]:

$$L = S - \varkappa' D_{S/\gamma} \qquad (1.5.3)$$

The value of the D correction is here assigned to the so-called reduced blackening, which is obtained by dividing the blackening by γ, instead of assigning it to the blackening value itself. There is, of course, only a formal analogy between the transformation coefficients $\varkappa$ and $\varkappa'$ of the $P_\varkappa$- and L-transformations, respectively.

Investigations have shown [37] that there is a linear relation between the coefficient, $\varkappa'$, of Eq. (1.5.3) and the blackening:

$$\varkappa' = \varkappa_0 \left(1 - \frac{S}{S_L}\right)$$

where $\varkappa_0$ denotes the value of $\varkappa'$ extrapolated to $S = 0$; S_L is the value of S extrapolated to $\varkappa' = 0$, and S_L is, at the same time, the blackening value belonging to the lowest point of the straight part of the characteristic curve.

When the value of $\varkappa'$ is substituted from the latter equation into Eq. (1.5.3), the L-transformation has the following form:

$$L = S - \varkappa_0 \left(1 - \frac{S}{S_L}\right) \cdot D_{S/\gamma}. \tag{1.5.4}$$

Let us bear in mind that for both the P_L- and L-transformations the range of values is the following:

$$0 < S < S_L.$$

The blackening values forming the straight part of the characteristic curve are identical with the transformed blackening values, i.e. when $S \geq S_L$, $P_L = S$ and $L \equiv S$.

The practical application of the L-transformation is much more simple than the use of the P_L-transformation. It has, however, the drawback that in order to apply it, one has to know two wavelength-dependent values ($\varkappa_0$ and S_L), in addition to the value of gradation. Further examinations have shown that the two transformation coefficients can be merged into a single one. Thereby an easily manageable equation is obtained: this is the l-transformation.

1.6 l-TRANSFORMATION

A rigorous study [38] of a great number of data relating to the two constants of the L-transformation described by Eq. (1.5.4) has shown that the values of $\varkappa_0$ and S_L are practically identical, in the cases of most different emulsions, wavelengths and methods of development:

$$\varkappa_0 = S_L.$$

Let us substitute this value of S_L into Eq. (1.5.4):

$$L = S - (\varkappa_0 - S) \cdot D_{S/\gamma}. \tag{1.6.1}$$

As the studies have shown [38] *the transformation coefficient $\varkappa_0$ becomes independent of the wavelength as well, when its value is divided by γ.*

Let us denote the values divided by γ by the corresponding small letters:

$$L = l, \quad \frac{S}{\gamma} = s$$

and let us introduce the following analogous denotations:

$$\frac{\varkappa_0}{\gamma} = \frac{S_L}{\gamma} = k, \quad D_{S/\gamma} = D_s = d. \tag{1.6.2}$$

Substituting these symbols into Eq. (1.6.1) we obtain the *equation of l-transformation* (Fig. 1.6.1):

$$\boxed{l = s - (k - s) \cdot d\,.} \tag{1.6.3}$$

The fundamental equation of the l-transformation using the photographic transmittancy, T_p, is the following:

$$l = \lg 10^k \cdot T_P^{1/\gamma} \cdot \lg (1 - T_P^{1/\gamma}) - \lg T_P^{1/\gamma}.$$

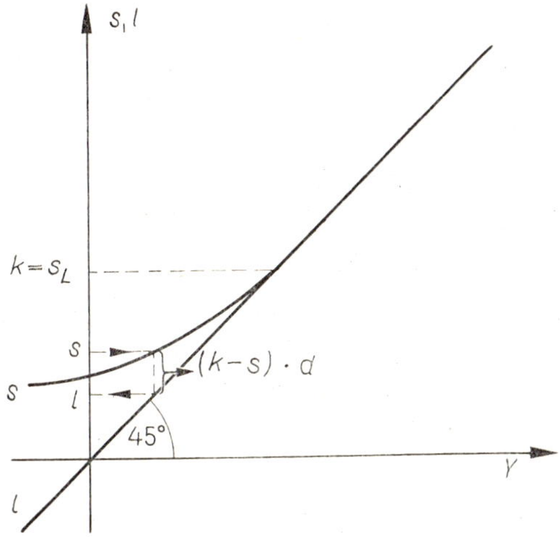

FIG. 1.6.1. The reduced blackening curve, s, and the corresponding l-transformed blackening curve

Naturally, l-transformation also relates only to the underexposed part of the characteristic curve. When $s \geq k$, $s = l$; i.e. only in the case $s < k$ is the transformation necessary. The value of d is a definite function of the reduced blackening. The l-transformation, therefore, contains only one constant, and this renders its practical application very convenient. This could be achieved owing to the fact that *k is a constant, independent of the blackening, wavelength and the gradation of the spectrum plate;* its value depends only on the type and age (way of storing) of the emulsion, and on the conditions of development [39].

A comparison with the P-transformation written in the appropriate form, i.e. obtained on division by γ, is then self-suggesting. It must not, however, be forgotten that this similarity is entirely formal, since the meanings of the quantities D and d, and $\varkappa$ and k, are different and the transformed curves also differ. On the other hand, one has to emphasize that the quantities found in the l-transformation do have physical meaning [40].

In the case of a quantitative analysis, the logarithmic intensity ratio, ΔY, belonging to the analytical line-pair, which has fundamental importance, is the same as the difference, Δl, between the l-transformed values of the blackening of the two members of the line-pair [41]:

$$\Delta l = \Delta Y. \tag{1.6.4}$$

Thus l-transformation gives directly the logarithm of the intensity ratio. It is an additional advantage that the $s \geq k$ values—which usually constitute the majority of the measured data in spectrograms of analytical samples—need not be transformed at all, since the straight parts of the reduced characteristic curve and of the transformed characteristic curve coincide.

A theoretical interpretation of the transformation equations is possible by studying the properties of the photographic emulsions [42], the initiation and growth of crystal centers produced by the irradiation, and the absorption conditions in thin and thick layers. It is not the purpose of this book to deal with these theories, thus only the relevant literature references are listed here [14, 15, 23, 28, 36, 40, 43–54].

2. PRACTICE

2.1 INTRODUCTION

This part deals with the practical problems of the evaluation of spectrograms. The steps of the l-transformation, the methods for the determination of the values of γ and k, the background correction and the concentration calculations will be described. A number of problems of the analysis which are not in direct connection with the application of the l-transformation will be disregarded. Consequently, with the exception of the most important construction problems of the analytical curve, many topics will not be dealt with in the present discussion; such as the factors influencing the form of the analytical curve (sparking off effect, interelement effect, parallel shift, etc.), special methods (addition method, etc.), control methods (standards, etalons, the use of standards, etc.), the mechanization and automation of the evaluation process (application of computers and special equipment, programming, etc.), and the treatment of the measurement data (error calculations, statistical methods, etc.). For a study of these topics the relevant handbooks, text-books and collections of practical examples should be consulted.

In order to facilitate the practical use of l-transformation, the l-values belonging to different s values have been calculated by an electronic computer for the range $s \leq k$, and grouped according to the practically occurring k-values. These pairs of values are tabulated in the order of decreasing blackening values [1].

The application of l-transformation is very simple [1, 38, 39]. First the value of γ is determined (Section 2.2). In contrast to the procedure used in the case of the W-, $P_{1/2}$- and $P_{\varkappa}$-transformations, where γ referred to the transformed characteristic curve, γ means here the slope of the straight part of the original characteristic curve.

There are a number of possible ways of determining the transformation coefficient k (Section 2.3). The value of k is characteristic of the type of the emulsion. For this reason, in the case of routine analyses, the same k-value can be used for any wavelength, provided the conditions of development are the same. It is therefore enough to determine carefully the k-value beforehand, once for a given set of analytical parameters, and this can be used as long as the type of the emulsion and the way of developing remain unchanged. In contrast to other transformation procedures, and to the determination of γ, there is no need for individual measurements of k in the case of different plates and analytical lines.

The steps of the l-transformation are the following when the values of γ and k are known:

1. The measured blackening values, S, are divided by γ : $s = S/\gamma$.

2. The l values belonging to the reduced blackening values, s, are found from the table belonging to the given k value. In the cases $s \geq k$, this step, i.e. the transformation itself, is omitted, as $l \equiv s$.

3. The difference of the values l_x and l_r obtained by means of the analytical line-pair is formed. This Δl is numerically identical to the value of ΔY (the logarithm of the intensity ratio of the analytical line-pair), and can be used directly for the construction of the analytical curve and for the determination of the sample concentration, respectively.

The l-transformation gives sufficiently accurate results over the full optical spectral range [55]. Its applicability for long wavelengths or high γ-values [56] should be especially emphasized. In the ultraviolet range the accuracy of the $P_\varkappa$-transformation is also satisfactory, but in the visible range it is not. The l-transformation therefore is a great help especially in the evaluation of visible range spectra [57, 58], but it can also be successfully used for the evaluation of X-ray spectra [59] and for the examination of the near- (photographic) infrared range [60].

2.2 DETERMINATION OF THE γ-VALUE

The determination of the γ-value is done in the usual way, by any method applicable for the construction of the characteristic curve [10, 61–65]. The prerequisite of a correct γ-determination is that intensity signal-pairs of known intensity ratio should be available [61, 66, 67]. These can be spectral lines of known intensity ratio or, more appropriately, spectral line sections divided into two or more parts in a known intensity ratio by means of a step-filter or step-sector. The first method is more difficult and less frequently used.

The most suitable procedure is to use spectral lines of two intensity grades. The two-step rotating sector method has been in use mainly in the United States, and the two-step filter is mostly applied in Europe. The value of the intensity ratio of the two steps is called *sector constant* or *filter constant*, respectively, and is denoted by q. In the sector method, this constant is provided by the ratio of the central angles (ω_b, ω_a) of the two sector openings (Fig. 2.2.1):

$$q = \omega_\text{b}/\omega_\text{a}.$$

The filter constant is the ratio of the intensities I_b and I_a passing through the two filter steps:

$$q = I_\text{b}/I_\text{a}.$$

The sector constant is independent of the wavelength. The filter constant is, however, not independent of the wavelength, not even in the case of the so-called neutral filter layers. Both values are usually given in percent, i.e. the hundredfold of the q value.

The negative logarithm of the sector or filter constant, that is, the *logarithmic sector-*, or *filter-constant*, is denoted by ΔY_m, as suggested by Kaiser:

$$\Delta Y_m = - \lg q$$

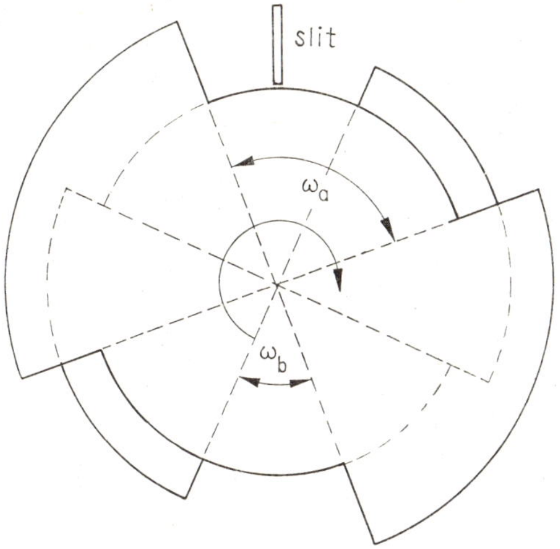

FIG. 2.2.1. The principle of the rotating two-step sector

Using appropriate lines of the spectra produced *by means of step-filters or step-sectors,* the blackening values are measured in the two steps, on the straight part of the characteristic curve. In order to increase accuracy, one has to use several measured data in the calculation. Using the same spectral line, the blackenings, S_b, measured on the less black part are deducted from the blackening values, S_a, measured on the blacker part. These differences, $\Delta S_{a,b}$, are the maximal blackening differences, ΔS_m, measured at the given wavelength for the same step filter or rotating sector. The $\Delta S_{a,b}$, values belonging to the same ΔY_m-values are maximal when both S_a and S_b are found on the straight part of the characteristic curve (see Fig. 1.4.1). Dividing $\overline{\Delta S_m}$, the arithmetical mean of the ΔS_m-values, by the logarithmic filter constant ΔY_m, we obtain the γ-value, as the following expression holds for the straight part of the characteristic curve [5]:

$$\gamma = \frac{\overline{\Delta S_m}}{\Delta Y_m}. \tag{2.2.1}$$

The error of the γ-determination is composed of the errors in the measurement of blackening and of the filter constant, according to Eq. (2.2.1). Most accurate are the procedures where γ is determined on the analytical line itself. An exception to this rule is the medium ultraviolet range. Between approximately 250 and 320 nm the γ-value is constant for most types of spectrum plates, therefore it is not necessary to determine the γ-value for each analytical line separately in this range [55].

Accordingly, the determination of the γ-value is a simple procedure. It is, however, necessary to know the precise value of the *filter constant* furthermore those blackening values which can be safely used in the calculations as they correspond to *the limits of the straight part of the characteristic curve.* Finally a sufficient number of *blackening value-pairs* are necessary for obtaining a true average.

2.2.1 DETERMINATION OF THE FILTER CONSTANT

The nominal filter constant value cannot be accepted as the true one in the case of commercial step-filters. That is only an approximate value for the filter constant and logarithmic filter constant. The nominal filter constant is generally given for white light, whereas *the actual value of the filter constant depends on the wavelength.* Recently manufacturers give certified filter constants in tabulated form for a number of wavelengths. These values are usually accurate, although precise periodic calibration of the transmittance of the filter is absolutely necessary because of ageing processes and contamination.

The method used for the calibration of a filter may considerably affect the dependability of the value of the filter constant. There are a number of methods for the calibration of the filter constant.

Calibration by *altering the distance of the light source from the slit* is exact, but cumbersome; it requires the use of an approximately point-like light source whose intensity does not fluctuate in time [67]. The usual arc and spark excitations are therefore unsatisfactory. On the other hand, if a continuously radiating source is used, the γ-value will be different from that obtained with a source radiating line spectra [68].

The calibration of a step-filter is also possible *by altering the aperture of the diaphragm placed in the way of the beam,* but in this case the filter constant must be exactly known at least for one wavelength [67].

The most general method is the application of a *spectrophotometer* for the determination of the filter constant. By means of a high-luminosity and sensitive photometer the step-filter can be simply calibrated for any wavelength. The filter should be placed in the way of the light beam of the spectrophotometer by means of a movable filter carrier having an appropriate cut. When the deflection of the spectrophotometer is set at 100% for the fully transmitting part of the filter, the value of the percentage transmittance or the ΔY_m-value can directly be obtained on the attenuated step by reading T% or the extinction. A correct γ-value is only obtained when the transmittance of the filter surfaces is completely uniform. This is important because during calibration a much larger filter surface must be used, owing to the intensity conditions, than the surface corresponding to the narrow slit width of the spectrograph used for obtaining the spectra.

The filter can also be calibrated on the spectrograph used for the analyses [69]. When a stable light source emitting continuous radiation is used, the characteristic curve is plotted for each step of the filter as a function of the logarithm of the slit width, *by varying the slit width.* The logarithm of the filter constant is equal to the horizontal distance measured on the abscissa, between the straight parts of the characteristic curves. This method is especially applicable when the filter layer is not placed on a flat surface, but on a curved one, e.g. on a lens. In such cases spectrophotometric calibration is not dependable.

The filter constant can be determined *by means of comparison with another filter of exactly known filter constant* [67, 70]. With the help of spectrograms made on the same plate by two (or more) filters, the gradation of the emulsion is determined using a filter with authentic filter constant, according to Eq. (2.2.1); the unknown filter constant is then calculated. When Eq. (2.2.1) is applied to the measured data obtained with both filters, there is, in fact, no need for the direct determination of the γ-value [55]:

$$\Delta Y_{m,x} = \frac{\overline{\Delta S_{m,x}}}{\overline{\Delta S_{m,r}}} \cdot \Delta Y_{m,r} \,.$$

In the above equation the indices x and r refer to the unknown and known filter constants, or to the blackening differences measured on the spectra with the aid of these filters, respectively. The advantage of the method is its simplicity; there is no need for special equipment (e.g. a stable source), only a calibrated step-filter is necessary. Any error in the filter constant of the latter naturally contributes to the error of the filter constant to be determined.

Whichever method is used for calibration, the ΔY_m-values are best plotted as a function of the wavelength on a diagram paper. The ΔY_m-values belonging to the analytical lines are read from this calibration graph.

2.2.2 THE STRAIGHT PART OF THE CHARACTERISTIC CURVE

It is necessary to know which blackening values belong to the straight part of the characteristic curve. The lower limit of the straight part, S_L, can be simply calculated from Eq. (1.6.2):

$$S_L = k \cdot \gamma \,.$$

The value of γ, S_L and the upper limit of the straight part of the characteristic curve increase with increasing wavelength, while k and ΔS_L are independent of the wavelength [70].

The length of the straight part of the characteristic curve, ΔS_L, is also independent of the type of the emulsion and the spectrograph. Its numerical value is influenced only by the conditions of development (the composition of the developer, the time and temperature of the development process) [71]. In the case of usual development conditions (e.g., Agfa 1 developer, 20°C, 5 minutes) *the value of ΔS_L can be taken as* **1.25 ± 0.07** [70].

The constancy of ΔS_L makes difficult the exact determination of high γ-values. The reason is that the filter constants of the two- and three-step filters provided for the spectrographs are chosen to fit the ultraviolet range. In the case of the frequently applied 50% transmittance two-step filters $\Delta Y_m \simeq$ **0.30.** From Eq. (2.2.1) it follows that

$$\Delta S_m = \gamma \cdot \Delta Y_m \,. \tag{2.2.2}$$

If therefore $\gamma = 1$, which frequently happens in the ultraviolet range, $\Delta S_m \cong$ **0.30.** When in the visible spectral range the γ-value is, e.g. **5** for some spectral line, then $\Delta S_m =$ **1.50,** i.e. it is a larger value than the full length of the straight part of the characteristic curve, ΔS_L. In this case, therefore, the values S_a and S_b cannot simultaneously be on the straight part of the characteristic curve. In the case of medium ΔY_m and high γ values, the γ-determination is impracticable [56]. If such measurement data were still used, the γ-values obtained would be lower than the real ones, since in this case smaller ΔS_m would be obtained than the actual value.

This difficulty can be surmounted by the use of the special, so-called 'γ-compensating selective two-step filter' of Nagy and Sámsoni [72]. This filter is made with a metal or carbon coating of appropriate thickness [73] whose transmittance increases with increasing wavelength. The logarithmic filter constant therefore decreases at a rate which corresponds to the normal increase of the gradations of emulsions normally used in spectral analysis [75].* As a consequence, according to Eq. (2.2.2), the value of ΔS_m is almost constant over a very wide spectral range.

2.2.3 BLACKENING VALUE PAIRS FOR THE DETERMINATION OF γ

It follows from what has been said above that the only further requirement for an accurate determination of γ is the availability of a sufficient number of blackening value pairs. In the medium ultraviolet spectral range this is no problem (Section 2.2), since here the blackening values of the lines neighbouring the analytical lines can be directly measured as well. At wavelengths longer than 320 nm, however, this procedure cannot be applied. In this range, first of all, the value of γ significantly varies with the increase of the wavelength, further, the γ-value measured on neighbouring lines fluctuates more than in the ultraviolet spectral range [55].

If the spectrum plates contain a great number of spectrograms, normally sufficient number of blackening value pairs can be found on the individual spectra [10]. However, it may happen that some plates do not contain suitable blackening value pairs in a certain wavelength range for the γ-determination. In such cases an iteration method can be applied for the l-transformation of low blackenings, provided the filter constant and the transformation coefficient are known [1, 38, 74], which is usually the case. The substitution of values for γ are tried until the difference Δl has become identical to ΔY_m. Since the k transformation coefficient does not depend on the wavelength, a k value determined at a different wavelength may also be used in this method.

One must also be careful to apply the same slit-width of the spectrograph for the determination of γ as is used for the analytical exposures, since the γ-value also depends on the slit-width [70, 74–78].

*In spectroscopy different emulsions are used. Process, half-tone or extra-hard plates are used to get fine grain and thin emulsion; they also give high γ- and higher λ-values. Faster emulsions all have coarser grains, or are 'sensitized' for special purposes, usually for some range of visible or infrared radiation

By the application of a two-step light attenuating device (filter, diaphragm, rotating sector or difference in the light path), pairs of spectra with different blackening values can also be produced by varying the exposure time. Owing to the reciprocity failure, or, in other words, as a result of the Schwartzschild effect, the amount of radiation reaching the emulsion is not a linear function of the exposition time, that is, the γ-values of the time- and intensity-scale characteristic curves are not identical. However, within the same spectrum plate, the exposure time is the same for both steps. Thus, the ratio of the exposures (amounts of radiation) is equal to the intensity ratio, *independently of the exposure time*, for any wavelength [79]. The γ value calculated this way gives the slope of the straight part of the characteristic curve having intensity-scale.

Thus, applying a two-step filter or sector, serial spectrograms are obtained on a plate using a metal giving a crowded spectrum (e.g. iron, nickel) or employing the material to be analyzed, if it is suitable in this respect; the spectrograms are made by varying the exposure time in some appropriate time scale; this series then can be used for obtaining a sufficient number of blackening value pairs for the determination of γ (and at the same time of k). The ratio of the subsequent exposure times varies generally between 1.2 and 1.6 depending on the type of emulsion, wavelength range and conditions of development.

2.3 METHODS FOR THE DETERMINATION OF k

The k constant of the l-transformation can be determined by different methods: by graphical procedure and calculation [23]. These methods differ in simplicity, speed and accuracy.

2.3.1 CALCULATION METHOD

The transformation coefficient k can be determined *by means of the fundamental equation* (1.6.3) *of the l-transformation*. On the analogy of the $P_\varkappa$-transformation, the difference of the transformed blackenings is written as follows:

$$\Delta l = \Delta s - k \cdot \Delta d + \Delta(sd). \tag{2.3.1}$$

The line blackenings are measured on the two steps of the two pairs of spectra made by some two-step attenuating device, preferably by a two-step filter. The transformed blackening difference Δl, on the left-hand side of Eq. (2.3.1) will be the same on the two spectra, since the filter constant gives rise to the same intensity ratio, and the value of γ is constant at the same wavelength.

Let us write Eq. (2.3.1) for the two pairs of spectra of different intensities. Denote the blackening values measured on the first and second pair by s' and s'', respectively; double differences are denoted by $\Delta\Delta$. Putting the right-hand sides equal (cf. Eq. (1.5.2)), the value of k can be calculated [11, 38]:

$$k = \frac{\Delta s' - \Delta s'' + \Delta(sd)' - \Delta(sd)''}{\Delta d' - \Delta d''} = \frac{\Delta\Delta s + \Delta\Delta(sd)}{\Delta\Delta d}.$$

This method is correct in principle; however, owing to the double difference formation, it is not accurate enough, similarly to the determination of the constant $\varkappa$ of the $P_\varkappa$-transformation (Section 1.5.4).

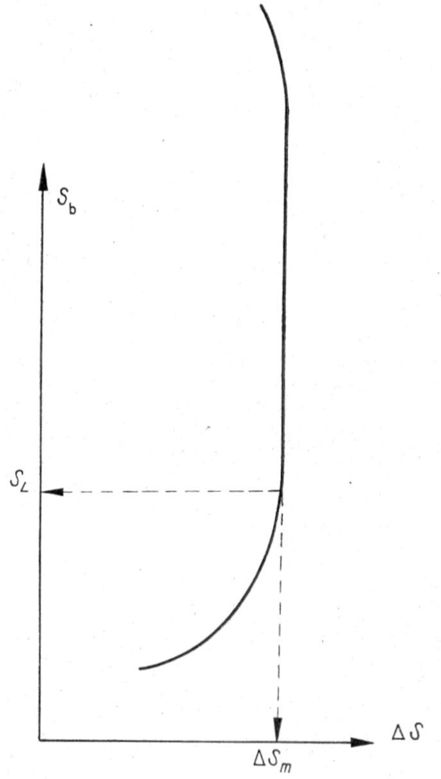

FIG. 2.3.1. The determination of S_L by means of the preliminary curve

2.3.2 USE OF THE CHARACTERISTIC CURVE

Whatever method was used for the construction of the *characteristic curve, the lowest blackening value on the straight part,* S_L, *can be directly read from it.* Dividing this by γ, the k-value is obtained according to Eq. (1.6.2).

2.3.3 METHOD OF USING PRELIMINARY CURVES

The value of k may be obtained without the construction of the characteristic curve, if *the method of using a preliminary curve* is applied [10].

Let us plot on the ordinate the blackenings, S_b, measured on the attenuated step, and on the abscissa the blackening differences between the two steps: $\Delta S = S_a - S_b$. The value of S_L can be read directly from the preliminary curve; this is the lowest S_b-value, where the curve which has been vertical for blackenings corresponding to the straight part, starts bending toward the ordinate axis [37]. Division of S_L by γ gives the value of k (Fig. 2.3.1).

2.3.4 DETERMINATION BY THE VALUE PAIRS S_a, S_b

The value of S_L and thereby k can be determined *directly,* without any actual construction, *from the blackening data* available for the construction of the characteristic curve. First the blackening pairs S_a, S_b are ordered according to decreasing value. The differences $\Delta S = S_a - S_b$ are formed and this value of S_b is equal to S_L at which the blackening difference, ΔS, begins to decrease. This is so because on the straight part the ΔS values scatter around the mean value, while if at least one of the blackening values (in this case S_b) lies on the lower, curved part of the characteristic curve, the ΔS values decrease.

The value of ΔS naturally decreases toward the high blackening values, too, if S_a lies on the *upper curved part* of the characteristic curve. This case has, however, no practical significance. It is only in the determination of γ where one must pay attention that the measured blackening values should not be higher than the upper limit of the straight part of the characteristic curve.

In these methods, which are directly or indirectly connected with the construction of the characteristic curve, the determination of k usually entails a larger scatter, as the value of S_L can thus be determined usually only to an accuracy of $\pm$**0.05–0.10** blackening units. *Therefore, the k-values obtained in this way can only be regarded as approximate orientation values.*

2.3.5 ITERATION METHOD

It is advisable to determine the k value for the plate types used *with high precision in advance.* To this end, *series of spectra are made by means of a two-step filter,* and possibly with *samples* of logarithmically *varying concentration* [10], or with *different exposure times* [61, 79]. The blackening values are measured at a number of wavelengths on the attenuated and on the non-attenuated steps of the spectra. The line blackening values are arranged in decreasing order. The values corresponding to the straight part of the characteristic curve are used for the γ-calculations (Section 2.2), while those representing the lower, curved part are employed for the calculation of k [38].

All blackening values are divided by γ. For the resulting reduced blackenings, s, the pertaining l-transformed blackenings are found from a table, for an appropriately chosen value of k. If one of the methods described in Section 2.3 has been used for the determination of the approximate value of k and so it is known in advance, this is used in the calculations. Otherwise, the approximate k values which may be expected are the following [41]: for extrahart type fresh plates of good quality: $k \cong$ **0.20.** For older 'extrahart' or fresh 'rapid' plates: $k \cong$ **0.30.** When the plate is of lower quality 'extrahart' or older 'rapid', the value of k is about **0.40.** For old, badly stored or low quality plates: $k \cong$ **0.50.**

The s values lower than the approximately known (or assumed) k values are transformed with the help of the table belonging to this k value, i.e. the pertaining values of l are found. For the reduced blackenings greater than k this procedure is omitted, as for these values $s = l$. Thereafter the transformed blackening differences ($l_a - l_b = \Delta l$) are calculated. If k was chosen correctly, the Δl values obtained for the correlated blackening pairs, or their average, will be identical with ΔY_m. When $\overline{\Delta l} < \Delta Y_m$, the transformation should be done again using a higher k-value, whereas if $\overline{\Delta l} > \Delta Y_m$, the repetition must be done with a lower k-value [1].

In order to speed up the iteration, the following rule should be noted: for not too low blackening values *a change of* $+$**0.01** *in* Δl *approximately corresponds to a change of* $\mp$(**0.02–0.03**) *in* k [38].

The iteration is continued, i.e. the transformation using a successively improved k-value is repeated until, with the exact k-value, $\overline{\Delta l} = \Delta Y_m$ has been obtained. With some practice and taking the above rule into account, the correct k-value is found within two or three steps.

The value of k can be determined most accurately by this method. This k can be used in the analyses as long as the type of the emulsion or the development conditions remain unchanged.

It has been mentioned that k is usually independent of the wavelength [39, 56]. The same average k-value can be used for calculations in the full spectral range [57]. Only emulsions sensitized for the visible range are exceptions to this rule. In the case of these plates usually a smaller k-value should be used *within the sensitized spectral range* than for the shorter wavelengths [39]. When, therefore, sensitized emulsions are used, the value of k is determined in two wavelength ranges.

2.4 BACKGROUND CORRECTION BY l-TRANSFORMATION

It is known that at low concentrations *the analytical curve deviates from the straight line if the background radiation is considerable*, or line coincidence occurs, or when the element to be determined is present in some of the materials or in the auxiliary electrode as a contamination. The most important literature data dealing with background correction are the works of Kaiser [17, 80], the references cited therein, and some other publications [76, 81].

The blackening measured on a line can arise from several kinds of radiation. The intensities are additive. There is, however, a logarithmic relationship between the radiation intensity and line blackening. As a consequence, *the blackenings are not additive*. The background blackening or the blackening of the coincident line cannot be subtracted from the total line blackening. First the intensities should be calculated from the total blackening and from the background blackening measured, e.g., beside the line. The corresponding intensity differences should then be recalculated to blackening, transformed blackening, or intensity logarithm [17], as required.

The general scheme of this procedure of background correction is the following:

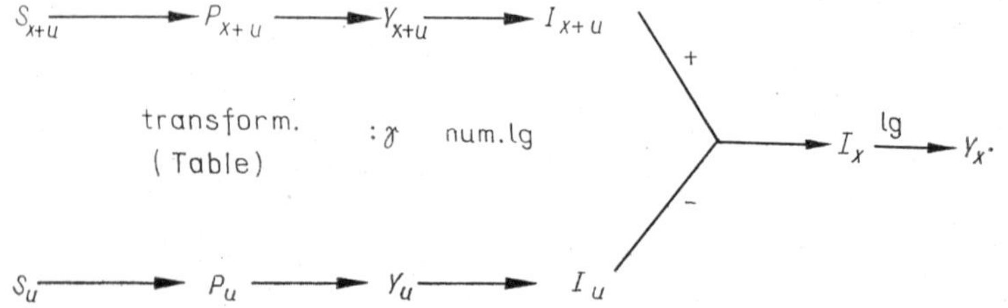

Instead of lengthy calculations, a simpler procedure is the use of Gaussian subtraction logarithm [82]. Since the l-transformed values are identical with the Y-values, the background correction can be simply applied:

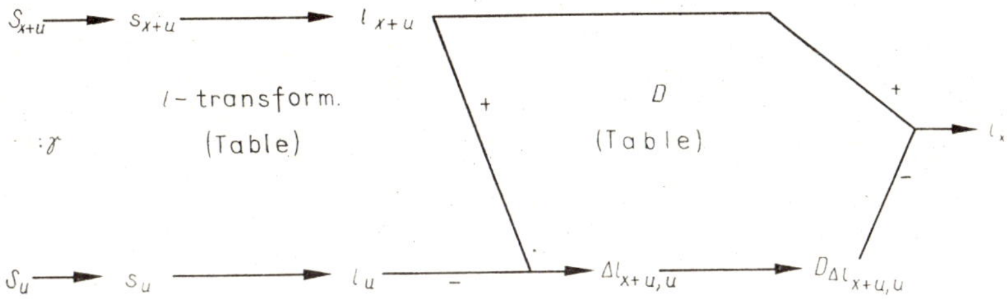

The S blackening values measured on the line and on the background are divided by γ, and from the l-transformed blackening, belonging to the resulting reduced line blackening, s, the transformed background blackening is subtracted. From Table 5.2 the value of D belonging to this difference is found and subtracted from the transformed blackening measured on the line. The result is the transformed blackening of the line corrected with the background, i.e. the logarithm of its intensity.

Essentially, the same procedure can be used for taking into account *line coincidence* and the *contamination of the auxiliary electrodes* [83].

2.5 DETERMINATION OF CONCENTRATION FROM THE Δl-VALUES

The first step of the determination of the concentration is *the construction of the analytical curve* (Fig. 1.3.1) by means of samples of known concentration. This curve is in most cases *linear* over a wide range of concentrations, under exactly defined experimental conditions.

The determination of the concentrations, c', corresponding to the $\Delta l \equiv \Delta Y$ values obtained by the analytical series of spectra, may be achieved *graphically or by means of calculation*. In both methods the parameters of the analytical curve, i.e. the values of c_r' and η according to Eq. (1.3.1), must be known. Instead of η – as suggested by Scheibe and Schöntag [84] – the relative concentration, c_m', belonging to ΔY_m can also be used (Fig. 2.5.1). The value of ΔY_m is required anyway, for the determination of γ. Occasional construction of the analytical curve by means of the c_r' and c_m' values is much simpler than the use of η instead of c_m'. The following relation exists among the quantities η, c_r' and c_m':

$$\eta = \frac{\lg c_m' - \lg c_r'}{\Delta Y_m} . \tag{2.5.1}$$

In the *graphical method* c' corresponding to Δl is read directly from the graph drawn on a logarithmic chart.

In the *calculation method* Δl is determined from Eq. (1.3.1). The value of η is calculated from Eq. (2.5.1).

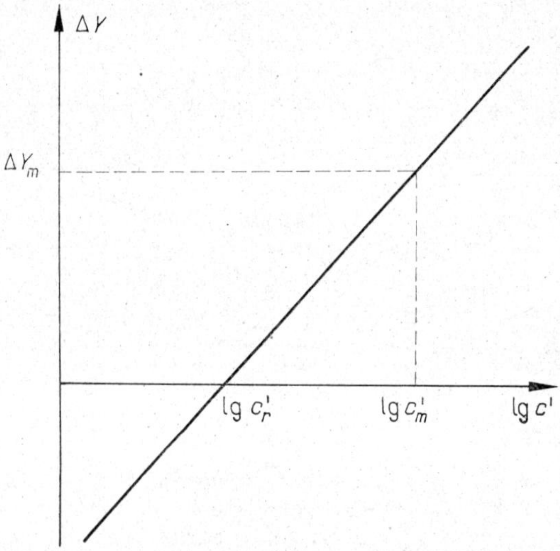

FIG. 2.5.1. The construction of the analytical curve by means of the parameters c'_r and c'_m

For the construction of the analytical curve *relative concentration* values are used. Thus the Δl-values obtained in the analyses always give relative concentration values, c'. What is needed in practice is the value expressed for a unit amount of the composite sample. The unit of the total amount of material can be, e.g. 1, 100, 10^6 or 10^9 g, the relative concentration being g/g, %, ppm or ppb, respectively. In the case of liquid materials, the unit amount of material (solution) can be, e.g., 1 ml or 100 ml. The quantity of the element in question calculated for this unit volume is usually given for example in μg or g. The concentration related to 100 weight- or volume unit is usually called weight percent and volume percent, respectively.

There are two cases in the calculation of concentration c_x from the relative concentration, c':

(a) *The* c_r *concentration of the reference element is constant* and identical in both the standard and the unknown sample.

(b) *The concentration of the reference element is not constant.*

In the first case the fundamental equation (Eq. (1.3.1)) of the evaluation can be written, utilizing the relation $c' = c_x/c_r$, as follows:

$$\lg c_x = \lg c_r + \lg c'_r + \eta \Delta l.$$

Since in this case not only c'_r but also c_r is constant, the analytical curve of the form $\lg c_x$, Δl can directly be constructed, and the c_x concentration values corresponding to the Δl-values can be read without any calculation.

In the examination of solutions and powders the *standard* is almost invariably given separately to the analytical sample, *in every case in the same quantity*. In such cases therefore c_r is always constant. It is also self-evident that if the value of c_r is practically (within $\mp 5\%$) *identical* in the standard and in the sample of unknown concentration, the above procedure can be applied directly. This is the case, e.g. with metals (aluminium, etc.) or different types of low alloyed steels (Fe $\geq 95\%$).

In case when the value of c_r cannot even practically be considered constant, the analytical curve must be constructed by using the relative concentrations, and the values of c' obtained from the Δl values must be converted into c_x values. This calculation can only be done if a so-called 'total analysis' is performed. This means that the relative concentrations of all elements have to be determined which are present at a significant concentration level in the sample. In this case the concentration to be determined is given by the following expression:

$$c_x = c' \frac{100}{100 + \sum_{1}^{n} c'_n}$$

where $\sum_{1}^{n} c'_n$ is the sum of the relative concentrations of the above-mentioned other elements.

3. NUMERICAL EXAMPLES

3.1 INTRODUCTION

In the following, actual examples will be used to illustrate the determination of the constants γ and k (Section 3.2) and the calculation of Δl (Section 3.3), from the simplest to the most complicated cases (Chapters 3.3.1–3.3.4). Finally, the concentration values corresponding to the Δl-values will be calculated as an example.

The calculations using the given measurement data have been described in word form. When necessary, sketches have been used for illustrating the steps of the calculations. At the end of each example the numerical data were collected from the initial data through the intermediate values up to the final results, in the form of tables that are easy to comprehend.

It must be emphasized that *the values of S, ΔS, s, Δs, l, Δl, ΔY_m and k are given in the numerical examples as the hundredfold of their original value.* This is done since usually the hundredfold values of the true blackenings are directly read from the S-scale of the densitometers. In this case at most values with one decimal digit number are used instead of three decimal digits, as this allows easy survey of the examples.

Accordingly, *the hundredfold values of the blackenings, transformed blackenings and logarithmic intensities are plotted on the ordinate of the analytical curves.* Obviously, if the concentration is determined by calculation and not by the graphical method, in other words, in the case of analyses without standards, *the algebraic values of the blackening (reducing and transformed blackening, logarithmic intensity, filter constant, transformation constant) must be used instead of the hundredfold values* in the equations, as it is done in the numerical examples.

In order to show clearly for all tabulated and numerical data which values are the original ones and which are multiplied by one hundred, *the original values are given in bold-faced type* (e.g., $\gamma = \mathbf{2.1}$), *whereas the hundredfolds are set in normal type* (e.g., $\Delta Y_m = 31.2$). No mistake is possible in the numerical values of the concentrations, which are always given in percent, therefore these values are given in usual type.

3.2 EXAMPLE FOR THE DETERMINATION OF γ AND k

In Sections 2.2 and 2.3 the methods for the calculation of γ and k were discussed. The following actual examples will illustrate the detailed calculations.

For the dependable calculation of γ and k one needs *the value of ΔY_m,* and about 15–20 of such *blackening value pairs,* S_a, S_b, which are approximately evenly distributed over the straight and the lower part of the char-

TABLE 3.2.1

Numerical Example for the Determination of γ and k

$$\Delta Y_m = 28.5$$

$$\overline{\Delta S_m} = \frac{\sum_1^n{}' \Delta S_m}{n} = \frac{364.0}{11} = 33.1$$

$$\gamma = \frac{\overline{\Delta S_m}}{\Delta Y_m} = \frac{33.1}{28.5} = \mathbf{1.16}$$

S_a	S_b	ΔS	$\gamma = \mathbf{1.16}$		$k = 30$			$k = 25$		
			s_a	s_b	l_a	l_b	Δl	l_a	l_b	Δl
167.0	134.0	33.0								
155.0	121.5	33.5								
142.5	108.7	33.8								
129.0	97.1	31.9								
117.2	84.2	33.0								
105.7	73.0	32.7								
96.8	63.3	33.5								
91.3	58.4	32.9								
83.6	50.0	33.6								
76.5	43.2	33.3								
65.2	32.4	32.8								
56.8	25.5	31.3	49.0	22.0	49.0	18.8	30.2	49.0	20.8	28.2
49.4	20.3	29.1	42.6	17.5	42.6	11.5	31.1	42.6	13.9	28.7
41.0	15.6	25.4	35.3	13.4	35.3	3.8	31.5	35.3	6.7	28.4
35.1	12.9	22.2	30.3	11.1	30.3	−1.2	31.5	30.3	2.1	28.2
31.4	10.9	20.5	27.1	9.4	26.1	−5.2	31.3	27.1	−1.7	28.6
25.7	8.1	17.6	22.2	7.0	19.1	−12.0	31.1	21.1	−7.9	29.0
22.6	6.7	15.9	19.5	5.8	14.8	−16.1	30.9	17.0	−11.5	28.5
19.0	5.2	13.8	16.4	4.5	9.6	−21.2	30.8	12.1	−16.1	28.2
16.2	4.1			$\overline{\Delta l} =$			31.0			28.5

acteristic curve. The blackening value pairs are produced by some device that brings about a known intensity *ratio,* usually by a step-filter or a rotating sector (Section 2.2).

In the present example such blackening value pairs are shown which were measured on an 'extrahart' type spectrum plate of medium fog level and produced by a spectral line chosen in the ultraviolet wavelength range. The blackening value pairs are listed in decreasing order (Table 3.2.1). The calibrated value 51.9% was obtained for the filter constant at the given wavelength for a two-step filter (nominal value: 50%) used for producing the blackening pair S_a, S_b. Thus the logarithmic filter constant is:

$$\Delta Y_m = 100 \cdot \lg \frac{100}{T\%} = 100 \cdot \lg \frac{100}{51.9} = 28.5 .$$

Columns 1 and 2 of Table 3.2.1 contain the (hundredfold) values of the blackenings S_a and S_b, as read on the densitometer. In the upper frame the (hundredfold) values of the filter constant are given. The blackenings $S \geq 5.0$ are used for the calculations, i.e. the data of the bottom line are not employed, since there $S_b < 5.0$.

First the differences $\Delta S = S_a - S_b$ (Column 3) are calculated. The limit above which *all* blackening values are found on the straight part of the characteristic curve, is given by a dotted line. This limit is established from the blackening differences in Column 3: at higher blackening values the ΔS values fluctuate around an average (in the present example this is about 33 blackening unit difference). As soon as S_b (and later, as the blackening values drop further, S_a as well) falls to the lower curved part, the value of ΔS will obviously begin to decrease. The dotted limiting line is drawn on this basis. The value of γ is determined from the blackening differences higher than this limit, while the lower blackening values are employed for the determination of the correct value of k, using the γ previously obtained. For the determination of k it is enough if only one of the employed blackening value (S_b) pairs is on the lower curved part.

The average of the blackening differences — in the present case: 11 — corresponding to the straight part is 33.1 (see the second frame). Accordingly, the value of γ is **1.16** from Eq. (2.2.1) (see third frame).

The value of γ should, of course, be determined for each plate and, within the wavelength range where γ changes even on the same plate separately for the analytical lines (Section 2.2).

The exact calculation of the value of k is started by the iteration method when the gradation of the characteristic curve is known [1]. The s_a and s_b-values, corresponding to S_a and S_b below the dotted line, are calculated, in other words, the blackening values are divided by γ (Columns 4 and 5).

Since the emulsion used was not fresh we selected the approximate value $k = 30$ (Section 2.3.5). From the table for $k = 30$ the l-transformed blackenings (l) belonging to the reduced blackenings (S) are read. These are given in Columns 6 and 7.

Naturally, the blackening values $s \geq k$ were not transformed, as for these values $l \equiv s$ (Section 1.6). Thus the four l_a-values in column 6 are identical with the first corresponding four s_a-values in Column 4 of Table 3.2.1. Thereafter the values $\Delta l = l_a - l_b$ (Column 8) are calculated and then averaged. The average $\overline{\Delta l}$ values are given in the lowest frame. Since the result obtained is $\Delta l = 31.0$, the choice of k was not correct, for in that case the calculations should have resulted in the transformed blackening difference of $\overline{\Delta l} = \Delta Y_m = 28.5$.

In the present case $\overline{\Delta l} - \Delta Y_m = 2.5$; this is the value by which the transformation gave a too high result in the case of $k = 30$. Applying the approximate rule concerning the relation between the variation of k and l (Section 2.3.5), the transformation is repeated with a k value smaller by 5–6 units.

Let us choose the value $k = 25$, and find the transformed blackening values l_a and l_b belonging to s_a and s_b in Columns 4 and 5, from the corresponding Table. The results are given in Columns 9 and 10, and the transformed blackening differences $\Delta l = l_a - l_b$ obtained for $k = 25$ are found in Column 11. (It must be remembered in the subtraction that the small l_b values are negative numbers!) The average of the $\overline{\Delta l}$-values is now 28.5, which is just equal to the expected value of $\Delta Y_m = 28.5$. The iteration therefore gave the correct result, $k = 25$, in the second step. It occurs very rarely in practice that two or three trials fail to give the correct result. It should be noted that if there is a difference of only a few tenths between the values of $\overline{\Delta l}$ and ΔY_m, the iteration may be discontinued. The remaining small difference will not give rise to any considerable error in practical analysis.

The value of k determined for a given type of emulsion can be used for plates treated under the same conditions of development, i.e. the k-value need not be determined in each case for every plate (Section 2.3.5).

3.3 EXAMPLES FOR THE CALCULATION OF THE Δl-VALUES

3.3.1 GENERAL REMARKS

When the constants are known, the determination is very simple. For the construction of the analytical curve, then for the calculation of the concentrations, the transformed blackening difference of the analytical line pair is required, since according to Eq. (1.6.4): $\Delta l = \Delta Y$.

3.3.2 CALCULATIONS WITHOUT BACKGROUND CORRECTION

In the simplest case the background blackening is so low that its value can be neglected in comparison with the measured blackening values. In such cases *background correction is unnecessary*. The calculation follows the scheme:

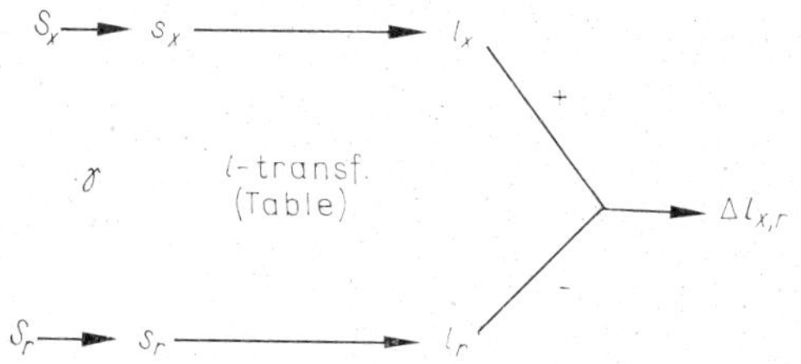

The signs + and – after l_x and l_r mean that the *lower* value should be subtracted from the *upper* one, without regard to the absolute value of l. In the following example, therefore, +39.8 should be deducted from –4.4.

Let the experimental data be the following: $S_x = 23.2$, $S_r = 89.6$, $\gamma = \mathbf{2.25}$ and $k = 32$. The calculations are shown in Table 3.3.2.1.

TABLE 3.3.2.1

	S	s	l	Δl
x	23.2	10.3	–4.4	–44.2
r	89.6	39.8	39.8	

3.3.3 CALCULATIONS WITH BACKGROUND CORRECTION FOR THE ANALYTICAL LINE

When the difference between the blackening of the analytical line and the background is small, *background correction* is necessary.

Let the following values apply: the value of the blackening measured on the analytical line (in the visible range): $S_{x+u} = 56.8$; the mean background blackening beside the line: $S_{u(x)} = 28.3$; the reference blackening: $S_{r+u} = 154.4$; the background blackening beside it: $S_{u(r)} = 16.0$; the gradation: $\gamma = \mathbf{4.15}$; and the transformation constant for the Gevaert Scientia 23 D 56 type spectrum plate: $k = 18$. As the value of the reference blackening did not change significantly on spectra made with different analytical samples, furthermore, since the difference between S_{r+u} and the not too high $S_{u(r)}$ is considerable, the blackening values measured on the line of the reference element need not be corrected for background intensity; S_{r+u} can be accepted to equal S_r.

Taking these into account, the l-transformation is effected according to the steps given in Eq. 2.4.1 (Table 3.3.3.1).

TABLE 3.3.3.1

	S	s	l	$\Delta l_{x+u,u}$	$D\Delta l_{x+u,u}$	l_x	$\Delta l_{x,r}$
$x+u$	56.8	13.7	11.2	13.8	56.5	−45.3	−82.5
u	28.3	6.8	−2.6				
r	154.5	37.2	37.2				

3.3.4 CALCULATIONS WITH BACKGROUND CORRECTION FOR THE ANALYTICAL LINE-PAIR

3.3.4.1 CALCULATIONS IN THE CASE $\gamma_x = \gamma_r = \gamma_{u(x)} = \gamma_{u(r)}$

If the background blackening level is very high, *the background correction must be performed on the blackening values of both members of the analytical line-pair*. In this case the background correction on the line of the reference element should be done in just the same way as on the analytical line.

The calculation scheme is the following:

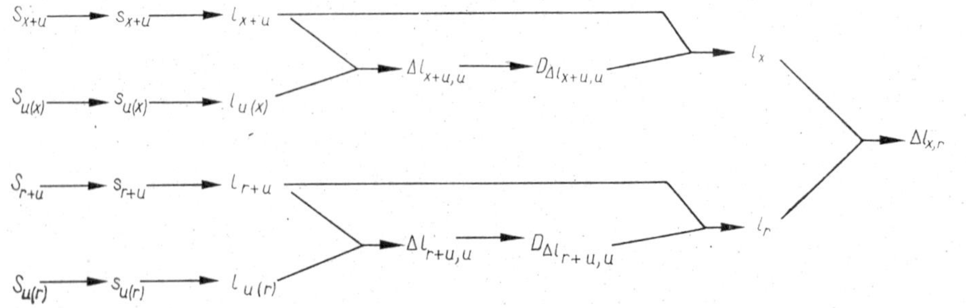

Let the blackening values measured on the analytical line-pair and in its neighbourhood in the spectral range around 360 nm on a plate with a high fog be the following:

$$S_{x+u} = 90.5,\ S_{u(x)} = 42.8,\ S_{r+u} = 70.3,\ S_{u(r)} = 37.2$$

and $\gamma = \mathbf{1.44}$; furthermore, corresponding to the γ value which is relatively low at this wavelength, the value of the transformation constant is high [39], $k = 55$. Of the tables given in this range for even k values, either of the neighbouring tables (in this case the one for $k = 54$ or $k = 56$) can be used. For exact calculations the required k values can be determined by *interpolation* from the two corresponding neighbouring tables, similarly as shown in the following examples.

Substitute the measured data into the steps of the former calculation scheme (Table 3.3.4.1).

TABLE 3.3.4.1

	S	s	l
$x + u(x)$	90.5	62.8	62.8
$u(x)$	42.8	29.7	21.9
$r + u(r)$	70.3	48.9	47.9
$u(r)$	37.2	25.8	15.6

40.9	21.5	41.3
Δl	$D_{\Delta l}$	l
32.3	28.0	19.9

$\Delta l = 21.4$

In this example the blackening measured at the analytical line need not have been transformed, i.e. $s_{x+u} = l_{x+u} = 62.8$, since this value is greater than $k = 55$. On the other hand, the reduced blackening of value 48.9 obtained by dividing the 70.3 line-blackening of the reference element by γ had to be transformed. It should be noted that the more usual case is when only the blackening values measured on the analytical line must be transformed.

3.3.4.2 CALCULATIONS IN THE CASE $\gamma_x = \gamma_{u(x)} \neq \gamma_r = \gamma_{u(r)}$

When the gradation calculated from the blackenings measured at the line of the reference element is different from the γ-value of the analytical line, then *different γ-values are used for the two members of the line-pair*. In the previous example, let the gradation measured at the analytical line be $\gamma_x = \mathbf{1.44}$, while at the somewhat higher wavelength of the reference element $\gamma_r = \mathbf{1.52}$. The calculations repeated with these values are contained in Table 3.3.4.2.

Table 3.3.4.2

	S	s	l
$x + u(x)$	90.5	62.8	62.8
$u(x)$	42.8	29.7	21.9
$r + u(r)$	70.3	46.3	44.7
$u(r)$	37.2	24.5	13.3

40.9	21.5	41.3
Δl	$D_{\Delta l}$	l
31.4	28.8	15.9

$\Delta l = 25.4$

It can be seen that a 5% change in the γ-value significantly affects the Δl-value at one of the analytical lines.

3.3.4.3 CALCULATIONS IN THE CASE $\gamma_x \neq \gamma_{u(x)} \neq \gamma_r \neq \gamma\, u_{u(r)}$

Examinations on the characteristic curve have shown that *different γ values are obtained for line spectra and for continuous spectra* [76, 81, 85]. The background gives a continuous spectrum, therefore the difference in the value of γ should be taken into account in accurate studies.

In the case of blackenings given in the previous example (using again the value $k = 55$), let the corresponding gradations be: $\gamma_{x+u(x)} = \mathbf{1.44}$, $\gamma_{u(x)} = \mathbf{1.30}$, $\gamma_{r+u(r)} = \mathbf{1.52}$, $\gamma_{u(r)} = \mathbf{1.37}$. Repeating the calculations with these γ values, the results given in Table 3.3.4.3 are obtained.

Table 3.3.4.3

	S	s	l
$x + u(x)$	90.5	62.8	62.8
$u(x)$	42.8	32.9	26.8
$r + u(r)$	70.3	46.3	44.7
$u(r)$	37.2	27.1	17.8

36.0	24.9	37.9
Δl	$D_{\Delta l}$	l
26.9	33.6	11.1

$\Delta l = 26.8$

Using separate γ-values for the calculations, the result obtained for the transformed blackening difference is other than the value obtained using a uniform γ.*

* When the calculations in Section 3.3.3 are repeated with the value of γ obtained for the continuous spectrum (when the reference densities were not corrected for background radiation) the result is —88.7 instead of $\Delta l = -82.5$, if the original $\gamma = \mathbf{4.15}$ is used for the analytical line and $\gamma = \mathbf{3.70}$ for the background blackening keeping $k = 18$ unchanged.

3.4 EXAMPLES FOR THE CALCULATION OF CONCENTRATION FROM THE Δl VALUES

In our examples the Δl transformed blackening differences belonging to the analytical line-pair were calculated from the measured blackening values. The Δl values give the logarithm of the intensity ratio of the analytical line-pairs, they can therefore be used directly for the calculations of the concentration.

From the value of Δl ($= \Delta Y$) calculated by the method described in the previous examples, the c' relative concentration can be obtained according to Eq. (1.3.1) when c'_r and η or, instead of the latter, the quantities c'_m and ΔY_m are known (Section 2.5). In this case an equation is used, therefore ΔY_m and Δl mean the *true values* of the filter constant and transformed blackening differences, respectively.

Let us use the following values in our next example:

$$c'_r = 0.355\%,\ c'_m = 0.79\% \text{ and } \Delta Y_m = \mathbf{0.285}.$$

The value of η, from Eq. (2.5.1) is the following:

$$\eta = \frac{\lg c'_m - \lg c'_r}{\Delta Y_m} = \frac{-0.102 + 0.450}{\mathbf{0.285}} = \mathbf{1.22}$$

TABLE 3.4.1

$100 \cdot \Delta l$	17.2	−36.8	4.9	−12.6	−0.5
$\lg c'$	−0.241	−0.899	−0.390	−0.604	−0.456
c' %	0.574	0.126	0.407	0.249	0.350

The Δl-values of five samples of unknown concentration, for the same element, are listed in the first line of Table 3.4.1. In order to prevent misunderstanding, it is explicitly given in the caption of the table that the values are hundredfolds of Δl, so that for further calculations they should be divided by 100. According to Eq. (1.3.1), at given values of the constants, when the Δl-values are successively substituted into the equation $\lg c' = 1.22\,\Delta l - 0.450$, the values of $\lg c'$ and the concentrations, c', are found in the second and third line of the table, respectively.

Let us construct by means of the above parameters the analytical curve on a logarithmic chart and read from it the c' values corresponding to the Δl-values found in Table 3.4.1. Compare these results with the calculated c'-values shown in Table 3.4.1.

The calculation of the concentration c_x from c' requires only a multiplication by a factor, so that no separate example is given for this calculation.

4. REFERENCES

1. TÖRÖK, T., ZIMMER, K.: *l*-Transformation Tables and their Application in Quantitative Spectral Analysis. Publication of the Scientific Society for Mechanical Engineering, Budapest (1964)
2. LOMAKIN, A.: Z. anorg. u. allg. Chem. *187*, 75 (1930)
3. SCHEIBE, G., SCHNETTLER, O.: Naturwiss. *19*, 134 (1931)
4. SCHEIBE, G., LINSTRÖM, C. F., SCHNETTLER, O.: Z. angew. Chem. *49*, 145 (1931)
5. KAISER, H.: Spectrochim. Acta *2*, 1 (1941)
6. GERLACH, W., SCHWEITZER, E.: Die chemische Spektralanalyse. Leopold Voss Verlag, Leipzig (1930)
7. KEREKES, I.: Magyar Kémiai Folyóirat *70*, 536 (1964)
8. KEREKES-CSÉTI, S., ÁG, Á.: Acta Chim. Hung. *41*, 281 (1964)
9. HURTER, F., DRIFFIELD, V. C.: J. Soc. Chem. Ind. *9*, 455 (1890)
10. TÖRÖK, T.: Photography in Quantitative Spectral Analysis without the Use of Reference Samples. Publication of the Institute for Special Engineering Studies, Budapest, V. *59*, II (1948)
    TÖRÖK, T.: Annales Univ. Sci. Budapest , Sect. Chim. *2*, 415 (1960)
    TÖRÖK, T.: Acta Geol. et Geogr. Univ. Comenianae, Bratislava, Geol. *6*, 122 (1959)
11. ZIMMER, K.: Rozpr. Narodn. Techn. Muzea, Praha, Řada Fis. Chem. *4*, 42 (1962)
12. VARSÁNYI, GY.: Candidate Thesis, Budapest (1954)
13. SAMPSON, R. A.: Mon. Not. Roy. Astron. Soc. *85*, 212 (1925)
14. BAKER, E. A.: Proc. Roy. Soc. Edinburgh *45*, 166 (1925)
15. BAKER, E. A.: Proc. Roy. Soc. Edinburgh *47*, 34 (1927)
16. SEIDEL, W.: Lecture at the Meeting of the Spectral Analytical Committee of the Metal and Ore Society, Düsseldorf, 21st July (1939)
17. HONERJÄGER-SOHM, M., KAISER, H.: Spectrochim. Acta *2*, 396 (1944)
18. TÖRÖK, T., ZIMMER, K.: Magyar Kémiai Folyóirat *66*, 477 (1960)
    TÖRÖK, T., ZIMMER, K.: Acta Chim. Hung. *29*, 273 (1961)
19. KAISER, H.: Spectrochim. Acta *3*, 159 (1947)
20. KAISER, H.: Spectrochim. Acta *4*, 351 (1951)
21. ZIMMER, K., TÖRÖK, T.: Acta Chim. Hung. *28*, 59 (1961)
    ZIMMER, K., TÖRÖK, T.: Magyar Kémiai Folyóirat *69*, 495 (1963)
22. BOUMANS, P. W. J. M.: Dissertation, Amsterdam, 1961
23. ZIMMER, K.: Emissionspektroskopie. Akademie Verlag, Berlin (1964), p. 46
    ZIMMER, K.: Notes of the lecture given at the Meeting of the Spectral Analytical Committee of the Scientific Society for Mechanical Engineering, 7th June (1963)
24. WADEWITZ, S.: Spectrochim. Acta *11*, 430 (1957)
25. WADEWITZ, S.: Acta Chim. Hung. *31*, 319 (1962)
26. HAFTKA, F. J.: Spectrochim. Acta *7*, 242 (1955)
27. TÖRÖK, T., ZENTAI, P.: Magyar Kémiai Folyóirat *66*, 480 (1960)
    TÖRÖK, T., ZENTAI, P.: Acta Chim. Hung. *30*, 11 (1962)
28. CANDLER, C.: Spectrochim. Acta *8*, 262 (1956)
29. HÜBNER, H. J., OHZU, H.: Z. angew. Phys. *11*, 460 (1959)
30. HÜBNER, H. J., OHZU, H.: VIII Coll. Spectr. Intern., Luzern (1959), H. R. Sauerländer and Co. Aarau (1960), p. 81
31. CROSSWHITE, H. M., DIEKE, G. H.: John Hopkins Univ. Bumblebee Series Report 202 (1953)

32. ARRAK, A.: Appl. Spectroscopy *11*, 38 (1957)
33. MORELLO, B.: Mikrochim. Acta *1955*, 390
34. MASI, O., MORELLO, B.: Spectrochim. Acta *5*, 130 (1952)
35. TÖRÖK, T.: Chemia Analyticzna, *7*, 47 (1962)
36. TÖRÖK, T.: Optik und Spektroskopie aller Wellenlängen. Akademie Verlag, Berlin (1962) p. 573
37. TÖRÖK, T., ZIMMER, K., RÉTI, S.: Mikrochim. Acta *(1962)* 611
TÖRÖK, T., ZIMMER, K., RÉTI, S.: Magyar Kémiai Folyóirat *68*, 395 (1962)
38. TÖRÖK, T., ZIMMER, K.: Acta Chim. Hung. *41*, 97 (1964)
TÖRÖK, T., ZIMMER, K.: Magyar Kémiai Folyóirat *70*, 358 (1964)
39. ZIMMER, K., TÖRÖK, T., ASZTALOS, I.: Chemia Analyticzna *11*, 1065 (1966)
ZIMMER, K., TÖRÖK, T., ASZTALOS, I.: Magyar Kémiai Folyóirat *72*, 417 (1966)
40. TÖRÖK, T.: Magyar Kémiai Folyóirat *69*, 269 (1963)
TÖRÖK, T.: Emissionspektroskopie. Akademie Verlag, Berlin (1964), p. 141
41. TÖRÖK, T., ZIMMER, K.: Annales Univ. Sci. Budapest., Sect. Chim. *8*, 11 (1966)
TÖRÖK, T., ZIMMER, K.: Lectures on the IXth Hungarian Meeting for Emission Spectral Analysis, Sopron, Publication of the Scientific Society for Mechanical Engineering, Budapest (1966)
42. MEES, C. E. K., JAMES, T. H.: The Theory of the Photographic Process. Macmillan Co., New York (1966)
43. BAKER, E. A.: Proc. Roy. Soc. Edinburgh *48*, 106 (1928)
44. BAKER, E. A.: J. Phot. Sci. *5*, 94 (1957)
45. PEJZULAEV, S. I.: Izvest. Akad. Nauk SSSR, Ser. Fiz. *18*, 273 (1954)
46. ARRAK, A.: Appl. Spectroscopy *12*, 171 (1958)
47. ARRAK, A.: Appl. Spectroscopy *13*, 143 (1959)
48. CANDLER, C.: Appl. Spectroscopy *13*, 17 (1959)
49. TÖRÖK, T.: Acta Chim. Hung. *41*, 155 (1964)
TÖRÖK, T.: Magyar Kémiai Folyóirat *70*, 278 (1964)
50. KÁNTOR, T.: Magyar Kémiai Folyóirat *70*, 541 (1964)
51. KÁNTOR, T.: Z. anal. Chem. *214*, 357 (1965)
52. TÖRÖK, T., DUDAR, E., HARSÁNYI, E.: Proc. XIII. Coll. Spectr. Intern., Ottawa (1967) 151
53. TÖRÖK, T.: Proc. XV Coll. Spectr. Intern. Plenary Conf., Madrid (1969) p. 171
54. TÖRÖK, T.: Proc. XIV Coll. Spectr. Intern., Debrecen (1967), p. 155
TÖRÖK, T.: Kémiai Közlemények *30*, 413 (1968)
55. ZIMMER, K., TÖRÖK, T., BIEDL, J.: Magyar Kémiai Folyóirat *72*, 425 (1966)
ZIMMER, K., TÖRÖK, T., BIEDL, J.: Acta Chim. Hung. *51*, 17 (1967)
56. ZIMMER, K., BASSA, ZS.: Proc. XIV. Coll. Spectr. Intern., Debrecen (1967), p. 847
ZIMMER, K., BASSA, ZS.: Magyar Kémiai Folyóirat *76*, 170 (1970)
57. ZIMMER, K., HAYPÁL, E.: Spectrochim. Acta, *24 B*, 655 (1969)
ZIMMER, K., HAYPÁL, E.: Magyar Kémiai Folyóirat, *76*, 83 (1970)
58. ZIMMER, K., IKRÉNYI, K.: Spectrochim. Acta, *25 B*, 425 (1970)
ZIMMER, K., IKRÉNYI, K.: Magyar Kémiai Folyóirat, *76*, 78 (1970)
59. MATHERNY, M.: Magyar Kémiai Folyóirat *71*, 88 (1965)
60. ZENTAI, P.: Acta Chim. Hung. *53*, 327 (1967)
61. ZIMMER, K.: Rev. Roum. de Chim. *10*, 707 (1965)
ZIMMER, K.: Gép *19*, 9 (1967)
62. CHURCHILL, J. R.: Ind. Eng. Chem., Anal. Ed. *16*, 653 (1944)
63. STRASHEIM, A.: Appl. Spectroscopy *12*, 137 (1958)
64. VAN CALKER, J., KASSELMANN, H.: Z. anal. Chem. *198*, 143 (1963)
65. GERBATSCH, R., BÄCHER, I.: Z. anal. Chem. *225*, 81 (1967)
66. HARRISON, G. R.: J. Opt. Soc. Amer. *24*, 59 (1934)
67. KAISER, H.: Spectrochim. Acta *3*, 518 (1949)
68. TÖRÖK, T., ZIMMER, K., BIEDL, J., ASZTALOS, I.: Acta Chim. Hung. *50*, 23 (1966)
TÖRÖK, T., ZIMMER, K., BIEDL, J., ASZTALOS, I.: Magyar Kémiai Folyóirat *72*, 430 (1966)
69. PLŠKO, E.: Magyar Kémiai Folyóirat *71*, 20 (1965)
70. ZIMMER, K.: Lectures on the XIth Hungarian Meeting for Emission Spectral Analysis, Dunaújváros, Publication of the Scientific Society for Mechanical Engineering (1968), p. 129

71. Zimmer, K., Tóth, B., Paksy, L., Gegus, E.: Lectures at the V Conference for Testing Materials, Budapest, Publication of the Scientific Society for Mechanical Engineering, (1970) p. 31
Zimmer, K., Tóth, B., Paksy, L., Gegus, E.: Magyar Kémiai Folyóirat, *77*, 200 (1971)
Zimmer, K., Tóth, B., Paksy, L., Gegus, E.: Acta Chim. Hung., in the press
72. Nagy, Z., Sámsoni, Z., Benkő, K.: Magyar Kémiai Folyóirat *70*, 549 (1964)
73. Nagy, Z.: Magyar Kémiai Folyóirat *73*, 104 (1967)
74. Zimmer, K.: Proc. XIV. Coll. Spectr. Intern., Debrecen (1967) p. 435
Zimmer, K.: Kémiai Közlemények *32*, 435 (1969)
75. Arrak, A.: Appl. Spectroscopy *16*, 124 (1962)
76. Slavin, M.: Appl. Spectroscopy *16*, 127 (1962)
77. Strasheim, A., de Villiers, D. B.: Acta Chim. Hung. *58*, 61 (1968)
78. Zimmer, K., Szabó, É.: Spectrochim. Acta *24 B*, 649 (1969)
Zimmer, K., Szabó, É.: Magyar Kémiai Folyóirat, *76*, 109 (1970)
79. Zimmer, K.: Magyar Kémiai Folyóirat *72*, 415 (1966)
80. Kaiser, H.: Spectrochim. Acta *3*, 297 (1948)
81. Zimmer, K., Szabó, É.: XV Coll. Spectr. Intern., Madrid (1969) IV, p. 55.
82. Preuss, E.: Z. angew. Mineral. *1*, 173 (1938)
83. Török, T.: Acta Chim. Hung. *49*, 11 (1966)
Török, T.: Magyar Kémiai Folyóirat *72*, 433 (1966)
84. Scheibe, G., Schöntag, A.: Arch. Eisenhüttenwesen *8*, 533 (1934–35)
85. Slavin, M.: Appl. Spectroscopy *16*, 173 (1962)

TABLE 5.1

l-values belonging to $s = S/\gamma$ for different *k*-values

The tables give the hundredfold of the real *k*-, *s*- and *l*-values

(For notations cf. Section 1.2, List of Symbols)

k=10
k=11
k=12
k=13

$k = 10$

s	*l*	*s*	*l*	*s*	*l*	*s*	*l*	*s*	*l*
10.0	10.0								
9.8	9.7	8.0	6.5	6.2	2.9	4.4	—1.3	2.6	— 6.5
9.6	9.3	7.8	6.1			4.2	—1.8	2.4	— 7.2
9.4	9.0	7.6	5.7	6.0	2.4			2.2	— 8.0
9.2	8.6	7.4	5.3	5.8	2.0	4.0	—2.3		
9.0	8.3	7.2	4.9	5.6	1.6	3.8	—2.9	2.0	— 8.8
8.8	7.9	7.0	4.5	5.4	1.1	3.6	—3.4	1.8	— 9.6
8.6	7.6	6.8	4.1	5.2	0.7	3.4	—4.0	1.6	—10.5
8.4	7.2	6.6	3.7	5.0	0.2	3.2	—4.6	1.4	—11.5
8.2	6.8	6.4	3.3	4.8	—0.3	3.0	—5.2	1.2	—12.6
				4.6	—0.8	2.8	—5.9	1.0	—13.8

$k = 11$

s	*l*	*s*	*l*	*s*	*l*	*s*	*l*	*s*	*l*
11.0	11.0								
10.8	10.7	9.0	7.5	7.0	3.7	5.0	—0.8	3.0	— 6.4
10.6	10.3	8.8	7.2	6.8	3.3	4.8	—1.3	2.8	— 7.1
10.4	10.0	8.6	6.8	6.6	2.9	4.6	—1.8	2.6	— 7.8
10.2	9.7	8.4	6.4	6.4	2.4	4.4	—2.3	2.4	— 8.5
10.0	9.3	8.2	6.1	6.2	2.0	4.2	—2.8	2.2	— 9.3
9.8	9.0	8.0	5.7	6.0	1.6	4.0	—3.4	2.0	—10.1
9.6	8.6	7.8	5.3	5.8	1.1	3.8	—4.0	1.8	—11.0
9.4	8.3	7.6	4.9	5.6	0.6	3.6	—4.5	1.6	—12.0
9.2	7.9	7.4	4.5	5.4	0.2	3.4	—5.1	1.4	—13.0
		7.2	4.1	5.2	—0.3	3.2	—5.8	1.2	—14.1
								1.0	—15.4

$k = 12$

s	l	s	l	s	l	s	l	s	l
12.0	12.0								
11.8	11.7	9.6	7.9	7.4	3.7	5.2	—1.2	3.0	— 7.6
11.6	11.3	9.4	7.6	7.2	3.3	5.0	—1.7	2.8	— 8.3
11.4	11.0	9.2	7.2	7.0	2.9	4.8	—2.3	2.6	— 9.0
11.2	10.7	9.0	6.8	6.8	2.4	4.6	—2.8	2.4	— 9.8
11.0	10.3	8.8	6.4	6.6	2.0	4.4	—3.3	2.2	—10.6
10.8	10.0	8.6	6.1	6.4	1.6	4.2	—3.9		
10.6	9.7	8.4	5.7	6.2	1.1			2.0	—11.5
10.4	9.3	8.2	5.3			4.0	—4.4	1.8	—12.4
10.2	9.0			6.0	0.7	3.8	—5.0	1.6	—13.4
		8.0	4.9	5.8	0.2	3.6	—5.6	1.4	—14.5
10.0	8.6	7.8	4.5	5.6	—0.3	3.4	—6.3	1.2	—15.7
9.8	8.3	7.6	4.1	5.4	—0.8	3.2	—6.9	1.0	—17.1

$k = 13$

s	l	s	l	s	l	s	l	s	l
13.0	13.0								
12.8	12.7	10.4	8.7	8.0	4.1	5.6	—1.2	3.2	— 8.1
12.6	12.4	10.2	8.3	7.8	3.7	5.4	—1.7		
12.4	12.0	10.0	7.9	7.6	3.3	5.2	—2.2	3.0	— 8.8
12.2	11.7	9.8	7.6	7.4	2.9			2.8	— 9.5
12.0	11.4	9.6	7.2	7.2	2.5	5.0	—2.7	2.6	—10.2
11.8	11.1	9.4	6.8			4.8	—3.2	2.4	—11.1
11.6	11.7	9.2	6.5	7.0	2.0	4.6	—3.8	2.2	—11.9
11.4	10.4			6.8	1.6	4.4	—4.3	2.0	—12.8
11.2	10.0	9.0	6.1	6.6	1.2	4.2	—4.9	1.8	—13.8
		8.8	5.7	6.4	0.7	4.0	—5.5	1.6	—14.8
11.0	9.7	8.6	5.3	6.2	0.2	3.8	—6.1	1.4	—16.0
10.8	9.4	8.4	4.9	6.0	—0.2	3.6	—6.7	1.2	—17.3
10.6	9.0	8.2	4.5	5.8	—0.7	3.4	—7.4	1.0	—18.7

$k = 14$

s	l	s	l	s	l	s	l	s	l
14.0	14.0								
13.8	13.7	11.2	9.4	8.6	4.6	6.0	—1.1	3.4	— 8.5
13.6	13.4	11.0	9.0	8.4	4.2	5.8	—1.6	3.2	— 9.2
13.4	13.1	10.8	8.7	8.2	3.8	5.6	—2.1	3.0	— 9.9
13.2	12.7	10.6	8.3			5.4	—2.6	2.8	—10.7
13.0	12.4	10.4	8.0	8.0	3.4	5.2	—3.1	2.6	—11.5
12.8	12.1	10.2	7.6	7.8	2.9	5.0	—3.7	2.4	—12.3
12.6	11.8			7.6	2.5	4.8	—4.2	2.2	—13.2
12.4	11.4	10.0	7.3	..4	2.1	4.6	—4.8		
12.2	11.1	9.8	6.9	7.2	1.7	4.4	—5.4	2.0	—14.2
		9.6	6.5	7.0	1.2	4.2	—5.9	1.8	—15.2
12.0	10.8	9.4	6.1	6.8	0.8			1.6	—16.3
11.8	10.4	9.2	5.7	6.6	0.3	4.0	—6.6	1.4	—17.5
11.6	10.1	9.0	5.4	6.4	—0.2	3.8	—7.2	1.2	—18.8
11.4	9.7	8.8	5.0	6.2	—0.6	3.6	—7.8	1.0	—20.4

$k = 15$

s	l	s	l	s	l	s	l	s	l
15.0	15.0								
14.8	14.7	12.0	10.1	9.2	5.0	6.6	—0.5	3.8	— 8.3
14.6	14.4	11.8	9.8			6.4	—1.0	3.6	— 8.9
14.4		11.6	9.5	9.0	4.6	6.2	—1.5	3.4	— 9.6
14.2	13.8	11.4	9.1	8.8	4.2	6.0	—2.0	3.2	—10.4
14.0	13.4	11.2	8.8	8.6	3.8	5.8	—2.5		
13.8	13.1			8.4	3.4	5.6	—3.0	3.0	—11.1
13.6	12.8	11.0	8.4	8.2	3.0	5.4	—3.5	2.8	—11.9
13.4	12.5	10.8	8.0	8.0	2.6	5.2	—4.1	2.6	—12.7
13.2	12.2	10.6	7.7	7.8	2.2			2.4	—13.6
		10.4	7.3	7.6	1.7	5.0	—4.6	2.2	—14.5
13.0	11.8	10.2	6.9	7.4	1.3	4.8	—5.2	2.0	—15.5
12.8	11.5	10.0	6.6	7.2	0.8	4.6	—5.8	1.8	—16.6
12.6	11.2	9.8	6.2			4.4	—6.4	1.6	—17.7
12.4	10.8	9.6	5.8	7.0	0.4	4.2	—7.0	1.4	—19.0
12.2	10.5	9.4	5.4	6.8	—0.1	4.0	—7.6	1.2	—20.4
								1.0	—22.0

$k = 16$

s	l	s	l	s	l	s	l	s	l
16.0	16.0								
15.8	15.7	13.0	11.2	10.0	5.9	7.0	—0.4	4.0	— 8.7
15.6	15.4	12.8	10.9	9.8	5.5	6.8	—0.9	3.8	— 9.3
15.4	15.1	12.6	10.6	9.6	5.1	6.6	—1.4	3.6	—10.0
15.2	14.8	12.4	10.2	9.4	4.7	6.4	—1.9	3.4	—10.8
15.0	14.5	12.2	9.9	9.2	4.3	6.2	—2.4	3.2	—11.5
14.8	14.2			9.0	3.9			3.0	—12.3
14.6	13.8	12.0	9.5	8.8	3.5	6.0	—2.9	2.8	—13.1
14.4	13.5	11.8	9.2	8.6	3.1	5.8	—3.4	2.6	—14.0
14.2	13.2	11.6	8.8	8.4	2.7	5.6	—3.9	2.4	—14.9
		11.4	8.5	8.2	2.2	5.4	—4.5	2.2	—15.8
14.0	12.9	11.2	8.1			5.2	—5.0		
13.8	12.6	11.0	7.7	8.0	1.8		—5.6	2.0	—16.9
13.6	12.2	10.8	7.4	7.8	[illegible]	4.8	—6.2	1.8	—18.0
13.4	11.9	10.6	7.0	7.6	0.9	4.6	—6.8	1.6	—19.2
13.2	11.6	10.4	6.6	7.4	0.5	4.4	—7.4	1.4	—20.5
		10.2	6.3	7.2	0.0	4.2	—8.0	1.2	—22.0
								1.0	—23.6

$k = 17$

s	l	s	l	s	l	s	l	s	l
17.0	17.0								
16.8	16.7	13.6	11.7	10.6	6.3	7.4	—0.3	4.4	— 8.4
16.6	16.4	13.4	11.3	10.4	6.0	7.2	—0.8	4.2	— 9.1
16.4	16.1	13.2	11.0	10.2	5.6			4.0	— 9.7
16.2	15.8			10.0	5.2	7.0	—1.3	3.8	—10.4
16.0	15.5	13.0	10.7	9.8	4.8	6.8	—1.8		—11.1
15.8	15.2	12.8	10.3	9.6	4.4	6.6	—2.2	3.4	—11.9
15.6	14.9	12.6	10.0	9.4	4.0	6.4	—2.8	3.2	—12.7
15.4	14.6	12.4	9.6	9.2	3.6	6.2	—3.3		
15.2	14.2	12.2	9.3			6.0	—3.8	3.0	—13.5
		12.0	8.9	9.0	3.2	5.8	—4.3	2.8	—14.3
15.0	13.9	11.8	8.6	8.8	2.8	5.6	—4.9	2.6	—15.2
14.8	13.6	11.6	8.2	8.6	2.3	5.4	—5.4	2.4	—16.1
14.6	13.3	11.4	7.8	8.4	1.9	5.2	—6.0	2.2	—17.1
14.4	13.0	11.2	7.5	8.2	1.5			2.0	—18.2
14.2	12.6			8.0	1.0	5.0	—6.6	1.8	—19.4
14.0	12.3	11.0	7.1	7.8	0.6	4.8	—7.2	1.6	—20.6
13.8	12.0	10.8	6.7	7.6	0.1	4.6	—7.8	1.4	—22.0
								1.2	—23.5
								1.0	—25.3

k = 18

s	*l*	*s*	*l*	*s*	*l*	*s*	*l*	*s*	*l*
18.0	18.0								
17.8	17.7	14.4	12.4	11.0	6.4	7.6	— 0.7		
17.6	17.4	14.2	12.1	10.8	6.1	7.4	— 1.1	4.0	—10.8
17.4	17.1			10.6	5.7	7.2	— 1.6	3.8	—11.5
17.2	16.8	14.0	11.8	10.4	5.3	7.0	— 2.1	3.6	—12.2
17.0	16.5	13.8	11.4	10.2	4.9	6.8	— 2.6	3.4	—13.0
16.8	16.2	13.6	11.1			6.6	— 3.1	3.2	—13.8
16.6	15.9	13.4	10.8	10.0	4.5	6.4	— 3.6	3.0	—14.6
16.4	15.6	13.2	10.4	9.8	4.1	6.2	— 4.1	2.8	—15.5
16.2	15.3	13.0	10.1	9.6	3.7			2.6	—16.4
		12.8	9.7	9.4	3.3	6.0	— 4.7	2.4	—17.4
16.0	15.0	12.6	9.4	9.2	2.9	5.8	— 5.2	2.2	—18.4
15.8	14.7	12.4	9.0	9.0	2.5	5.6	— 5.8		
15.6	14.4	12.2	8.7	8.8	2.0	5.4	— 6.3	2.0	—19.5
15.4	14.0			8.6	1.6	5.2	— 6.9	1.8	—20.7
15.2	13.7	12.0	8.3	8.4	1.2	5.0	— 7.5	1.6	—22.0
15.0	13.4	11.8	7.9	8.2	0.7	4.8	— 8.1	1.4	—23.5
14.8	13.1	11.6	7.6			4.6	— 8.8	1.2	—25.1
14.6	12.7	11.4	7.2	8.0	0.3	4.4	— 9.4	1.0	—26.9
		11.2	6.8	7.8	—0.2	4.2	—10.1		

k = 19

s	*l*	*s*	*l*	*s*	*l*	*s*	*l*	*s*	*l*
19.0	19.0								
18.8	18.7	17.0	16.0	15.0	12.9	13.0	9.5	11.0	5.8
18.6	18.4	16.8	15.7	14.8	12.5	12.8	9.1	10.8	5.4
18.4	18.1	16.6	15.4	14.6	12.2	12.6	8.8	10.6	5.0
18.2	17.8	16.4	15.1	14.4	11.9	12.4	8.4	10.4	4.6
18.0	17.5	16.2	14.8	14.2	11.5	12.2	8.0	10.2	4.2
17.8	17.2	16.0	14.5	14.0	11.2	12.0	7.7	10.0	3.8
17.6	16.9	15.8	14.1	13.8	10.9	11.8	7.3	9.8	3.4
17.4	16.6	15.6	13.8	13.6	10.5	11.6	6.9	9.6	3.0
17.2	16.3	15.4	13.5	13.4	10.2	11.4	6.6	9.4	2.6
		15.2	13.2	13.2	9.8	11.2	6.2	9.2	2.2

k = 19 *(continued)*

k = 19									
s	*l*	*s*	*l*	*s*	*l*	*s*	*l*	*s*	*l*
9.0	1.7	7.2	−2.4	5.6	− 6.7	4.0	−11.8	2.4	−18.7
8.8	1.3			5.4	− 7.3	3.8	−12.6	2.2	−19.7
8.6	0.8	7.0	−2.9	5.2	− 7.9	3.6	−13.3	2.0	−20.9
8.4	0.4	6.8	−3.4			3.4	−14.1	1.8	−22.1
8.2	0.1	6.6	−4.0	5.0	− 8.5	3.2	−14.9	1.6	−23.5
8.0	−0.5	6.4	−4.5	4.8	− 9.1	3.0	−15.8	1.4	−25.0
7.8	−1.0	6.2	−5.0	4.6	− 9.8	2.8	−16.7	1.2	−26.7
7.6	−1.5	6.0	−5.6	4.4	−10.4	2.6	−17.7	1.0	−28.6
7.4	−1.9	5.8	−6.1	4.2	−11.1				

k = 20									
s	*l*	*s*	*l*	*s*	*l*	*s*	*l*	*s*	*l*
20.0	20.0								
19.8	19.7	16.0	14.0	12.0	7.1	8.0	− 1.3	4.0	−12.9
19.6	19.4	15.8	13.6	11.8	6.7	7.8	− 1.8	3.8	−13.6
19.4	19.1	15.6	13.3	11.6	6.3	7.6	− 2.3	3.6	−14.4
19.2	18.8	15.4	13.0	11.4	5.9	7.4	− 2.7	3.4	−15.2
19.0	18.5	15.2	12.7	11.2	5.5	7.2	− 3.2	3.2	−16.1
18.8	18.3	15.0	12.3	11.0	5.1	7.0	− 3.8	3.0	−17.0
18.6	18.0	14.8	12.0	10.8	4.8	6.8	− 4.3	2.8	−17.9
18.4	17.7	14.6	11.7	10.6	4.4	6.6	− 4.8	2.6	−18.9
18.2	17.4	14.4	11.3	10.4	4.0	6.4	− 5.3	2.4	−19.9
		14.2	11.0	10.2	3.5	6.2	− 5.9	2.2	−21.1
18.0	17.1								
17.8	16.8	14.0	10.6	10.0	3.1	6.0	− 6.5	2.0	−22.2
17.6	16.5	13.8	10.3	9.8	2.7	5.8	− 7.0	1.8	−23.5
17.4	16.1	13.6	9.9	9.6	2.3	5.6	− 7.6	1.6	−24.9
17.2	15.8	13.4	9.6	9.4	1.9	5.4	− 8.2	1.4	−26.5
17.0	15.5	13.2	9.2	9.2	1.4	5.2	− 8.8	1.2	−28.2
16.8	15.2	13.0	8.9	9.0	1.0	5.0	− 9.5	1.0	−30.2
16.6	14.9	12.8	8.5	8.8	0.6	4.8	−10.1		
16.4	14.6	12.6	8.2	8.6	0.1	4.6	−10.8		
16.2	14.3	12.4	7.8	8.4	−0.4	4.4	−11.5		
		12.2	7.4	8.2	−0.8	4.2	−12.2		

$k = 21$									
s	*l*	*s*	*l*	*s*	*l*	*s*	*l*	*s*	*l*
21.0	21.0								
20.8	20.7	17.0	15.0	13.0	8.3	9.0	0.3	5.0	—10.4
20.6	20.4	16.8	14.7	12.8	7.9	8.8	—0.2	4.8	—11.1
20.4	20.1	16.6	14.4	12.6	7.6	8.6	—0.6	4.6	—11.8
20.2	19.9	16.4	14.1	12.4	7.2	8.4	—1.1	4.4	—12.5
20.0	19.6	16.2	13.8	12.2	6.8	8.2	—1.6	4.2	—13.2
19.8	19.3	16.0	13.4	12.0	6.4	8.0	—2.1	4.0	—13.9
19.6	19.0	15.8	13.1	11.8	6.1	7.8	—2.6	3.8	—14.7
19.4	18.7	15.6	12.8	11.6	5.7	7.6	—3.0	3.6	—15.5
19.2	18.4	15.4	12.5	11.4	5.3	7.4	—3.5	3.4	—16.4
		15.2	12.1	11.2	4.9	7.2	—4.1	3.2	—17.2
19.0	18.1								
18.8	17.8	15.0	11.8	11.0	4.5	7.0	—4.6	3.0	—18.2
18.6	17.5	14.8	11.5	10.8	4.1	6.8	—5.1	2.8	—19.1
18.4	17.2	14.6	11.1	10.6	3.7	6.6	—5.7	2.6	—20.1
18.2	16.9	14.4	10.8	10.4	3.3	6.4	—6.2	2.4	—21.2
18.0	16.6	14.2	10.4	10.2	2.9	6.2	—6.8	2.2	—22.4
17.8	16.3	14.0	10.1	10.0	2.4	6.0	—7.3	2.0	—23.6
17.6	16.0	13.8	9.7	9.8	2.0	5.8	—7.9	1.8	—24.9
17.4	15.7	13.6	9.4	9.6	1.6	5.6	—8.5	1.6	—26.4
17.2	15.9	13.4	9.0	9.4	1.2	5.4	—9.1	1.4	—28.0
		13.2	8.7	9.2	0.7	5.2	—9.8	1.2	—29.8
								1.0	—31.9

$k = 22$									
s	*l*	*s*	*l*	*s*	*l*	*s*	*l*	*s*	*l*
22.0	22.0								
21.8	21.7	17.6	15.5	13.4	8.4	9.2	0.0	5.0	—11.4
21.6	21.4	17.4	15.2	13.2	8.1	9.0	— 0.5	4.8	—12.1
21.4	21.2	17.2	14.9	13.0	7.7	8.8	— 0.9	4.6	—12.8
21.2	20.9	17.0	14.6	12.8	7.3	8.6	— 1.4	4.4	—13.5
21.0	20.6	16.8	14.2	12.6	7.0	8.4	— 1.9	4.2	—14.2
20.8	20.3	16.6	13.9	12.4	6.6	8.2	— 2.3		
20.6	20.0	16.4	13.6	12.2	6.2			4.0	—15.0
20.4	19.7	16.2	13.3			8.0	— 2.8	3.8	—15.8
20.2	19.4			12.0	5.8	7.8	— 3.3	3.6	—16.6
		16.0	12.9	11.8	5.4	7.6	— 3.8	3.4	—17.5
20.0	19.1	15.8	12.6	11.6	5.0	7.4	— 4.4	3.2	—18.4
19.8	18.8	15.6	12.3	11.4	4.7	7.2	— 4.9	3.0	—19.3
19.6	18.5	15.4	11.9	11.2	4.3	7.0	— 5.4	2.8	—20.3
19.4	18.2	15.2	11.6	11.0	3.8	6.8	— 6.0	2.6	—21.4
19.2	17.9	15.0	11.3	10.8	3.4	6.6	— 6.5	2.4	—22.5
19.0	17.6	14.8	10.9	10.6	3.0	6.4	— 7.1	2.2	—23.7
18.8	17.3	14.6	10.6	10.4	2.6	6.2	— 7.6		
18.6	17.0	14.4	10.2	10.2	2.2			2.0	—24.9
18.4	16.7	14.2	9.9			6.0	— 8.2	1.8	—26.3
18.2	16.4			10.0	1.8	5.8	— 8.8	1.6	—27.8
		14.0	9.5	9.8	1.3	5.6	— 9.4	1.4	—29.5
18.0	16.1	13.8	9.2	9.6	0.9	5.4	—10.1	1.2	—31.3
17.8	15.8	13.6	8.8	9.4	0.4	5.2	—10.7	1.0	—33.5

$k = 23$

s	*l*	*s*	*l*	*s*	*l*	*s*	*l*	*s*	*l*
23.0	23.0								
22.8	22.7	18.4	16.3	14.0	9.0	9.6	0.2	5.2	—11.7
22.6	22.4	18.2	16.0	13.8	8.6	9.4	— 0.3		
22.4	22.2	18.0	15.7	13.6	8.2	9.2	— 0.7	5.0	—12.3
22.2	21.9	17.8	15.3	13.4	7.9			4.8	—13.0
22.0	21.6	17.6	15.0	13.2	7.5	9.0	— 1.2	4.6	—13.8
21.8	21.3	17.4	14.7			8.8	— 1.7	4.4	—14.5
21.6	21.0	17.2	14.4	13.0	7.1	8.6	— 2.1	4.2	—15.3
21.4	20.7			12.8	6.8	8.4	— 2.6	4.0	—16.1
21.2	20.5	17.0	14.1	12.6	6.4	8.2	— 3.1	3.8	—16.9
		16.8	13.7	12.4	6.0	8.0	— 3.6	3.6	—17.7
21.0	20.2	16.6	13.4	12.2	5.6	7.8	— 4.1	3.4	—18.6
20.8	19.9	16.4	13.1	12.0	5.2	7.6	— 4.6	3.2	—19.5
20.6	19.6	16.2	12.8	11.8	4.8	7.4	— 5.2		
20.4	19.3	16.0	12.4	11.6	4.4	7.2	— 5.7	3.0	—20.5
20.2	19.0	15.8	12.1	11.4	4.0			2.8	—21.5
20.0	18.7	15.6	11.7	11.2	3.6	7.0	— 6.2	2.6	—22.6
19.8	18.4	15.4	11.4			6.8	— 6.8	2.4	—23.8
19.6	18.1	15.2	11.1	11.0	3.2	6.6	— 7.4	2.2	—25.0
19.4	17.8			10.8	2.8	6.4	— 7.9	2.0	—26.3
19.2	17.5	15.0	10.7	10.6	2.4	6.2	— 8.5	1.8	—27.7
		14.8	10.4	10.4	2.0	6.0	— 9.1	1.6	—29.3
19.0	17.2	14.6	10.0	10.2	1.5	5.8	— 9.7	1.4	—31.0
18.8	16.9	14.4	9.7	10.0	1.1	5.6	—10.4	1.2	—32.9
18.6	16.6	14.2	9.3	9.8	0.6	5.4	—11.0	1.0	—35.1

$k = 24$									
s	*l*	*s*	*l*	*s*	*l*	*s*	*l*	*s*	*l*
24.0	24.0								
23.8	23.7	19.2	17.1	14.6	9.5	10.0	— 0.4	5.4	—11.9
23.6	23.4	19.0	16.7	14.4	9.1	9.8	— 0.1	5.2	—12.6
23.4	23.2	18.8	16.4	14.2	8.8	9.6	— 0.5	5.0	—13.3
23.2	22.9	18.6	16.1			9.4	— 1.0	4.8	—14.0
23.0	22.6	18.4	15.8	14.0	8.4	9.2	— 1.4	4.6	—14.8
22.8	22.3	18.2	15.5	13.8	8.0	9.0	— 1.9	4.4	—15.5
22.6	22.1			13.6	7.7	8.8	— 2.4	4.2	—16.3
22.4	21.8	18.0	15.2	13.4	7.3	8.6	— 2.9		
22.2	21.5	17.8	14.9	13.2	6.9	8.4	— 3.4	4.0	—17.1
		17.6	14.5	13.0	6.5	8.2	— 3.9	3.8	—18.0
22.0	21.2	17.4	14.2	12.8	6.2			3.6	—18.8
21.8	20.9	17.2	13.9	12.6	5.8	8.0	— 4.4	3.4	—19.7
21.6	20.6	17.0	13.6	12.4	5.4	7.8	— 4.9	3.2	—20.7
21.4	20.3	16.8	13.2	12.2	5.0	7.6	— 5.4	3.0	—21.7
21.2	20.0	16.6	12.9			7.4	— 6.0	2.8	—22.7
21.0	19.8	16.4	12.6	12.0	4.6	7.2	— 6.5	2.6	—23.8
20.8	19.5	16.2	12.2	11.8	4.2	7.0	— 7.1	2.4	—25.0
20.6	19.2			11.6	3.9	6.8	— 7.6	2.2	—26.3
20.4	18.9	16.0	11.9	11.4	3.4	6.6	— 8.2		
20.2	18.6	15.8	11.6	11.2	3.0	6.4	— 8.8	2.0	—27.6
		15.6	11.2	11.0	2.5	6.2	— 9.4	1.8	—29.1
20.0	18.3	15.4	10.9	10.8	2.1			1.6	—30.7
19.8	18.0	15.2	10.5	10.6	1.7	6.0	—10.0	1.4	—32.5
19.6	17.7	15.0	10.2	10.4	1.3	5.8	—10.6	1.2	—34.5
19.4	17.4	14.8	9.8	10.2	0.8	5.6	—11.3	1.0	—36.8

$k = 25$

s	*l*	*s*	*l*	*s*	*l*	*s*	*l*	*s*	*l*
25.0	25.0								
24.8	24.7	20.0	17.8	15.2	10.0	10.6	1.0	5.8	−11.5
24.6	24.5	19.8	17.5			10.4	0.6	5.6	−12.2
24.4	24.2	19.6	17.2	15.0	9.7	10.2	0.1	5.4	−12.9
24.2	23.9	19.4	16.9	14.8	9.3	10.0	− 0.3	5.2	−13.6
24.0	23.6	19.2	16.6	14.6	8.9	9.8	− 0.8		
23.8	23.4			14.4	8.6	9.6	− 1.2	5.0	−14.3
23.6	23.1	19.0	16.3	14.2	8.2	9.4	− 1.7	4.8	−15.0
23.4	22.8	18.8	16.0	14.0	7.8	9.2	− 2.2	4.6	−15.8
23.2	22.5	18.6	15.7	13.8	7.5			4.4	−16.5
		18.4	15.4	13.6	7.1	9.0	− 2.6	4.2	−17.3
23.0	22.2	18.2	15.0	13.4	6.7	8.8	− 3.1	4.0	−18.2
22.8	21.9	18.0	14.7	13.2	6.3	8.6	− 3.6	3.8	−19.0
22.6	21.7	17.8	14.4			8.4	− 4.1	3.6	−19.9
22.4	21.4	17.6	14.1	13.0	6.0	8.2	− 4.6	3.4	−20.9
22.2	21.1	17.4	13.7	12.8	5.6	8.0	− 5.2	3.2	−21.8
22.0	20.8	17.2	13.4	12.6	5.2	7.8	− 5.7		
21.8	20.5			12.4	4.8	7.6	− 6.2	3.0	−22.9
21.6	20.2	17.0	13.1	12.2	4.4	7.4	− 6.8	2.8	−23.9
21.4	19.9	16.8	12.8	12.0	4.0	7.2	− 7.3	2.6	−25.1
21.2	19.6	16.6	12.4	11.8	3.6			2.4	−26.3
		16.4	12.1	11.6	3.2	7.0	− 7.9	2.2	−27.6
21.0	19.3	16.2	11.7	11.4	2.7	6.8	− 8.5	2.0	−29.0
20.8	19.0	16.0	11.4	11.2	2.3	6.6	− 9.1	1.8	−30.5
20.6	18.7	15.8	11.1			6.4	− 9.7	1.6	−32.1
20.4	18.4	15.6	10.7	11.0	1.9	6.2	−10.3	1.4	−34.0
20.2	18.1	15.4	10.4	10.8	1.5	6.0	−10.9	1.2	−36.0
								1.0	−38.4

$k = 26$									
s	*l*	*s*	*l*	*s*	*l*	*s*	*l*	*s*	*l*
26.0	26.0								
25.8	25.7	20.8	18.6	16.0	10.9	11.0	1.2	6.0	−11.8
25.6	25.5	20.6	18.3	15.8	10.5	10.8	0.8	5.8	−12.4
25.4	25.2	20.4	18.0	15.6	10.2	10.6	0.4	5.6	−13.1
25.2	24.9	20.2	17.7	15.4	9.8	10.4	0.2	5.4	−13.8
25.0	24.6			15.2	9.5	10.2	− 0.5	5.2	−14.5
24.8	24.4	20.0	17.4	15.0	9.1			5.0	−15.2
24.6	24.1	19.8	17.1	14.8	8.8	10.0	− 1.0	4.8	−16.0
24.4	23.8	19.6	16.8	14.6	8.4	9.8	− 1.5	4.6	−16.8
24.2	23.5	19.4	16.5	14.4	8.0	9.6	− 1.9	4.4	−17.5
		19.2	16.2	14.2	7.7	9.4	− 2.4	4.2	−18.4
24.0	23.3	19.0	15.8			9.2	− 2.9		
23.8	23.0	18.8	15.5	14.0	7.3	9.0	− 3.4	4.0	−19.2
23.6	22.7	18.6	15.2	13.8	6.9	8.8	− 3.9	3.8	−20.1
23.4	22.4	18.4	14.9	13.6	6.5	8.6	− 4.4	3.6	−21.0
23.2	22.1	18.2	14.6	13.4	6.1	8.4	− 4.9	3.4	−22.0
23.0	21.8			13.2	5.8	8.2	− 5.4	3.2	−23.0
22.8	21.6	18.0	14.2	13.0	5.4			3.0	−24.0
22.6	21.3	17.8	13.9	12.8	5.0	8.0	− 5.9	2.8	−25.1
22.4	21.0	17.6	13.6	12.6	4.6	7.8	− 6.5	2.6	−26.3
22.2	20.7	17.4	13.3	12.4	4.2	7.6	− 7.0	2.4	−27.6
		17.2	12.9	12.2	3.8	7.4	− 7.6	2.2	−28.9
22.0	20.4	17.0	12.6			7.2	− 8.1		
21.8	20.1	16.8	12.3	12.0	3.4	7.0	− 8.7	2.0	−30.3
21.6	19.8	16.6	11.9	11.8	2.9	6.8	− 9.3	1.8	−31.9
21.4	19.5	16.4	11.6	11.6	2.5	6.6	− 9.9	1.6	−33.6
21.2	19.2	16.2	11.2	11.4	2.1	6.4	−10.5	1.4	−35.5
21.0	18.9			11.2	1.7	6.2	−11.1	1.2	−37.6
								1.0	−40.1

k = 27									
s	*l*	*s*	*l*	*s*	*l*	*s*	*l*	*s*	*l*
27.0	27.0								
26.8	26.7	21.6	19.4	16.4	11.1	11.2	1.0	6.0	−12.7
26.6	26.5	21.4	19.1	16.2	10.7			5.8	−13.3
26.4	26.2	21.2	18.8	16.0	10.4	11.0	0.6	5.6	−14.0
26.2	25.9			15.8	10.0	10.8	0.2	5.4	−14.7
26.0	25.7	21.0	18.5	15.6	9.7	10.6	− 0.3	5.2	−15.5
25.8	25.4	20.8	18.2	15.4	9.3	10.4	− 0.8		
25.6	25.1	20.6	17.9	15.2	8.9	10.2	− 1.2	5.0	−16.2
25.4	24.8	20.4	17.6			10.0	− 1.7	4.8	−17.0
25.2	24.6	20.2	17.3	15.0	8.6	9.8	− 2.1	4.6	−17.8
		20.0	17.0	14.8	8.2	9.6	− 2.6	4.4	−18.6
25.0	24.3	19.8	16.7	14.6	7.8	9.4	− 3.1	4.2	−19.4
24.8	24.0	19.6	16.3	14.4	7.5	9.2	− 3.6	4.0	−20.3
24.6	23.7	19.4	16.0	14.2	7.1			3.8	−21.2
24.4	23.4	19.2	15.7	14.0	6.7	9.0	− 4.1	3.6	−22.1
24.2	23.2			13.8	6.3	8.8	− 4.6	3.4	−23.1
24.0	22.9	19.0	15.4	13.6	6.0	8.6	− 5.1	3.2	−24.1
23.8	22.6	18.8	15.1	13.4	5.6	8.4	− 5.6		
23.6	22.3	18.6	14.8	13.2	5.2	8.2	− 6.2	3.0	−25.2
23.4	22.0	18.4	14.4			8.0	− 6.7	2.8	−26.4
23.2	21.7	18.2	14.1	13.0	4.8	7.8	− 7.3	2.6	−27.6
		18.0	13.8	12.8	4.4	7.6	− 7.8	2.4	−28.8
23.0	21.5	17.8	13.4	12.6	4.0	7.4	− 8.4	2.2	−30.2
22.8	21.2	17.6	13.1	12.4	3.6	7.2	− 9.0	2.0	−31.7
22.6	20.9	17.4	12.8	12.2	3.2			1.8	−33.3
22.4	20.6	17.2	12.4	12.0	2.7	7.0	− 9.5	1.6	−35.0
22.2	20.3			11.8	2.3	6.8	−10.1	1.4	−37.0
22.0	20.0	17.0	12.1	11.6	1.9	6.6	−10.8	1.2	−39.2
21.8	19.7	16.8	11.8	11.4	1.5	6.4	−11.4	1.0	−41.7
		16.6	11.4			6.2	−12.0		

$k = 28$									
s	*l*	*s*	*l*	*s*	*l*	*s*	*l*	*s*	*l*
28.0	28.0								
27.8	27.7	22.4	20.2	17.0	11.6	11.6	1.3	6.2	—12.9
27.6	27.5	22.2	19.9	16.8	11.3	11.4	0.9		
27.4	27.2			16.6	10.9	11.2	0.4	6.0	—13.6
27.2	26.9	22.0	19.6	16.4	10.6	11.0	— 0.1	5.8	—14.2
27.0	26.7	21.8	19.3	16.2	10.2	10.8	— 0.5	5.6	—14.9
26.8	26.4	21.6	19.0			10.6	— 1.0	5.4	—15.7
26.6	26.1	21.4	18.7	16.0	9.9	10.4	— 1.4	5.2	—16.4
26.4	25.9	21.2	18.4	15.8	9.5	10.2	— 1.9	5.0	—17.2
26.2	25.6	21.0	18.1	15.6	9.1			4.8	—17.9
		20.8	17.8	15.4	8.8	10.0	— 2.4	4.6	—18.7
26.0	25.3	20.6	17.5	15.2	8.4	9.8	— 2.8	4.4	—19.6
25.8	25.0	20.4	17.2	15.0	8.1	9.6	— 3.3	4.2	—20.4
25.6	24.8	20.2	16.9	14.8	7.7	9.4	— 3.8		
25.4	24.5			14.6	7.3	9.2	— 4.3	4.0	—21.3
25.2	24.2	20.0	16.5	14.4	7.0	9.0	— 4.8	3.8	—22.3
25.0	23.9	19.8	16.2	14.2	6.5	8.8	— 5.3	3.6	—23.2
24.8	23.6	19.6	15.9			8.6	— 5.9	3.4	—24.2
24.6	23.4	19.4	15.6	14.0	6.2	8.4	— 6.4	3.2	—25.3
24.4	23.1	19.2	15.3	13.8	5.8	8.2	— 6.9	3.0	—26.4
24.2	22.8	19.0	14.9	13.6	5.4			2.8	—27.6
		18.8	14.6	13.4	5.0	8.0	— 7.5	2.6	—28.8
24.0	22.5	18.6	14.3	13.2	4.6	7.8	— 8.0	2.4	—30.1
23.8	22.2	18.4	14.0	13.0	4.2	7.6	— 8.6	2.2	—31.5
23.6	21.9	18.2	13.7	12.8	3.8	7.4	— 9.2		
23.4	21.7			12.6	3.4	7.2	— 9.8	2.0	—33.0
23.2	21.4	18.0	13.3	12.4	3.0	7.0	—10.4	1.8	—34.7
23.0	21.1	17.8	13.0	12.2	2.5	6.8	—11.0	1.6	—36.5
22.8	20.8	17.6	12.6			6.6	—11.6	1.4	—38.5
22.6	20.5	17.4	12.3	12.0	2.1	6.4	—12.2	1.2	—40.7
		17.2	12.0	11.8	1.7			1.0	—43.4

$k = 29$

s	*l*	*s*	*l*	*s*	*l*	*s*	*l*	*s*	*l*
29.0	29.0								
28.8	28.7	23.2	21.0	17.6	12.2	12.0	1.5	6.4	—13.1
28.6	28.5			17.4	11.8	11.8	1.1	6.2	—13.8
28.4	28.2	23.0	20.7	17.2	11.5	11.6	0.6	6.0	—14.5
28.2	27.9	22.8	20.4			11.4	0.2	5.8	—15.2
28.0	27.7	22.6	20.1	17.0	11.1	11.2	0.3	5.6	—15.9
27.8	27.4	22.4	19.8	16.8	10.8			5.4	—16.6
27.6	27.1	22.2	19.5	16.6	10.4	11.0	— 0.7	5.2	—17.4
27.4	26.9	22.0	19.2	16.4	10.1	10.8	— 1.2		
27.2	26.6	21.8	18.9	16.2	9.7	10.6	— 1.6	5.0	—18.1
		21.6	18.6	16.0	9.4	10.4	— 2.1	4.8	—18.9
27.0	26.3	21.4	18.3	15.8	9.0	10.2	— 2.6	4.6	—19.7
26.8	26.1	21.2	18.0	15.6	8.6	10.0	— 3.0	4.4	—20.6
26.6	25.8			15.4	8.3	9.8	— 3.5	4.2	—21.5
26.4	25.5	21.0	17.7	15.2	7.9	9.6	— 4.0	4.0	—22.4
26.2	25.2	20.8	17.4			9.4	— 4.5	3.8	—23.3
26.0	25.0	20.6	17.0	15.0	7.5	9.2	— 5.0	3.6	—24.3
25.8	24.7	20.4	16.7	14.8	7.1			3.4	—25.4
25.6	24.4	20.2	16.4	14.6	6.8	9.0	— 5.5	3.2	—26.4
25.4	24.1	20.0	16.1	14.4	6.4	8.8	— 6.1		
25.2	23.8	19.8	15.8	14.2	6.0	8.6	— 6.6	3.0	—27.6
		19.6	15.5	14.0	5.6	8.4	— 7.1	2.8	—28.8
25.0	23.6	19.4	15.1	13.8	5.2	8.2	— 7.7	2.6	—30.0
24.8	23.3	19.2	14.8	13.6	4.8	8.0	— 8.2	2.4	—31.4
24.6	23.0			13.4	4.4	7.8	— 8.8	2.2	—32.8
24.4	22.7	19.0	14.5	13.2	4.0	7.6	— 9.4	2.0	—34.4
24.2	22.4	18.8	14.2			7.4	—10.0	1.8	—36.0
24.0	22.1	18.6	13.8	13.0	3.6	7.2	—10.6	1.6	—37.9
23.8	21.9	18.4	13.5	12.8	3.2			1.4	—40.0
23.6	21.6	18.2	13.2	12.6	2.8	7.0	—11.2	1.2	—42.3
23.4	21.3	18.0	12.8	12.4	2.4	6.8	—11.8	1.0	—45.0
		17.8	12.5	12.2	1.9	6.6	—12.5		

$k = 30$									
s	*l*	*s*	*l*	*s*	*l*	*s*	*l*	*s*	*l*
30.0	30.0								
29.8	29.7	24.0	21.8	18.2	12.7	12.4	1.8	6.6	—13.3
29.6	29.5	23.8	21.5			12.2	1.3	6.4	—14.0
29.4	29.2	23.6	21.2	18.0	12.4			6.2	—14.6
29.2	29.0	23.4	20.9	17.8	12.0	12.0	0.9		
29.0	28.7	23.2	20.6	17.6	11.7	11.8	0.5	6.0	—15.3
28.8	28.4	23.0	20.3	17.4	11.3	11.6	0.0	5.8	—16.1
28.6	28.2	22.8	20.0	17.2	11.0	11.4	— 0.4	5.6	—16.8
28.4	27.9	22.6	19.7	17.0	10.6	11.2	— 0.9	5.4	—17.5
28.2	27.6	22.4	19.4	16.8	10.3	11.0	— 1.4	5.2	—18.3
		22.2	19.1	16.6	9.9	10.8	— 1.8	5.0	—19.1
28.0	27.4			16.4	9.6	10.6	— 2.3	4.8	—19.9
27.8	27.1	22.0	18.8	16.2	9.2	10.4	— 2.8	4.6	—20.7
27.6	26.8	21.8	18.5			10.2	— 3.2	4.4	—21.6
27.4	26.5	21.6	18.2	16.0	8.8			4.2	—22.5
27.2	26.3	21.4	17.9	15.8	8.5	10.0	— 3.7		
27.0	26.0	21.2	17.6	15.6	8.1	9.8	— 4.2	4.0	—23.4
26.8	25.7	21.0	17.3	15.4	7.7	9.6	— 4.7	3.8	—24.4
26.6	25.4	20.8	16.9	15.2	7.4	9.4	— 5.2	3.6	—25.4
26.4	25.2	20.6	16.6	15.0	7.0	9.2	— 5.8	3.4	—26.5
26.2	24.9	20.4	16.3	14.8	6.6	9.0	— 6.3	3.2	—27.6
		20.2	16.0	14.6	6.2	8.8	— 6.8	3.0	—28.7
26.0	24.6			14.4	5.8	8.6	— 7.4	2.8	—30.0
25.8	24.3	20.0	15.7	14.2	5.4	8.4	— 7.9	2.6	—31.3
25.6	24.1	19.8	15.3			8.2	— 8.5	2.4	—32.6
25.4	23.8	19.6	15.0	14.0	5.0			2.2	—34.1
25.2	23.5	19.4	14.7	13.8	4.6	8.0	— 9.0		
25.0	23.2	19.2	14.4	13.6	4.2	7.8	— 9.6	2.0	—35.7
24.8	22.9	19.0	14.0	13.4	3.8	7.6	—10.2	1.8	—37.4
24.6	22.6	18.8	13.7	13.2	3.4	7.4	—10.8	1.6	—39.3
24.4	22.3	18.6	13.4	13.0	3.0	7.2	—11.4	1.4	—41.5
24.2	22.1	18.4	13.0	12.8	2.6	7.0	—12.0	1.2	—43.9
				12.6	2.2	6.8	—12.7	1.0	—46.6

$k = 31$									
s	*l*	*s*	*l*	*s*	*l*	*s*	*l*	*s*	*l*
31.0	31.0								
30.8	30.7	25.0	22.8	19.0	13.6	13.0	2.4	7.0	—12.9
30.6	30.5	24.8	22.6	18.8	13.3	12.8	2.0	6.8	—13.5
30.4	30.2	24.6	22.3	18.6	12.9	12.6	1.6	6.6	—14.2
30.2	30.0	24.4	22.0	18.4	12.6	12.4	1.1	6.4	—14.8
30.0	29.7	24.2	21.7	18.2	12.2	12.2	0.7	6.2	—15.5
29.8	29.4	24.0	21.4	18.0	11.9	12.0	0.3	6.0	—16.2
29.6	29.2	23.8	21.1	17.8	11.6	11.8	— 0.2	5.8	—17.0
29.4	28.9	23.6	20.8	17.6	11.2	11.6	— 0.6	5.6	—17.7
29.2	28.6	23.4	20.5	17.4	10.9	11.4	— 1.1	5.4	—18.5
		23.2	20.2	17.2	10.5	11.2	— 1.5	5.2	—19.2
29.0	28.4								
28.8	28.1	23.0	19.9	17.0	10.1	11.0	— 2.0	5.0	—20.1
28.6	27.8	22.8	19.6	16.8	9.8	10.8	— 2.5	4.8	—20.9
28.4	27.6	22.6	19.3	16.6	9.4	10.6	— 3.0	4.6	—21.7
28.2	27.3	22.4	19.0	16.4	9.1	10.4	— 3.4	4.4	—22.6
28.0	27.0	22.2	18.7	16.2	8.7	10.2	— 3.9	4.2	—23.5
27.8	26.8	22.0	18.4	16.0	8.3	10.0	— 4.4	4.0	—24.5
27.6	26.5	21.8	18.1	15.8	8.0	9.8	— 4.9	3.8	—25.5
27.4	26.2	21.6	17.8	15.6	7.6	9.6	— 5.4	3.6	—26.5
27.2	25.9	21.4	17.5	15.4	7.2	9.4	— 6.0	3.4	—27.6
		21.2	17.2	15.2	6.8	9.2	— 6.5	3.2	—28.7
27.0	25.7								
26.8	25.4	21.0	16.8	15.0	6.4	9.0	— 7.0	3.0	—29.9
26.6	25.1	20.8	16.5	14.8	6.1	8.8	— 7.6	2.8	—31.2
26.4	24.8	20.6	16.2	14.6	5.7	8.6	— 8.1	2.6	—32.5
26.2	24.5	20.4	15.9	14.4	5.3	8.4	— 8.7	2.4	—33.9
26.0	24.3	20.2	15.6	14.2	4.9	8.2	— 9.2	2.2	—35.4
25.8	24.0	20.0	15.2	14.0	4.5	8.0	— 9.8	2.0	—37.1
25.6	23.7	19.8	14.9	13.8	4.1	7.8	—10.4	1.8	—38.8
25.4	23.4	19.6	14.6	13.6	3.7	7.6	—11.0	1.6	—40.8
25.2	23.1	19.4	14.3	13.4	3.3	7.4	—11.6	1.4	—43.0
		19.2	13.9	13.2	2.8	7.2	—12.2	1.2	—45.4
								1.0	—48.3

$k = 32$									
s	*l*	*s*	*l*	*s*	*l*	*s*	*l*	*s*	*l*
32.0	32.0								
31.8	31.7	25.6	23.4	19.4	13.8	13.2	2.3	7.0	—13.7
31.6	31.5	25.4	23.1	19.2	13.5	13.0	1.8	6.8	—14.3
31.4	31.2	25.2	22.8	19.0	13.1	12.8	1.4	6.6	—15.0
31.2	31.0	25.0	22.5	18.8	12.8	12.6	1.0	6.4	—15.7
31.0	30.7	24.8	22.2	18.6	12.5	12.4	0.5	6.2	—16.4
30.8	30.4	24.6	21.9	18.4	12.1	12.2	0.1		
30.6	30.2	24.4	21.6	18.2	11.8			6.0	—17.1
30.4	29.9	24.2	21.3			12.0	— 0.3	5.8	—17.9
30.2	29.7			18.0	11.4	11.8	— 0.8	5.6	—18.6
		24.0	21.0	17.8	11.1	11.6	— 1.3	5.4	—19.4
30.0	29.4	23.8	20.7	17.6	10.7	11.4	— 1.7	5.2	—20.2
29.8	29.1	23.6	20.4	17.4	10.4	11.2	— 2.2	5.0	—21.0
29.6	28.9	23.4	20.1	17.2	10.0	11.0	— 2.7	4.8	—21.9
29.4	28.6	23.2	19.8	17.0	9.7	10.8	— 3.1	4.6	—22.7
29.2	28.3	23.0	19.5	16.8	9.3	10.6	— 3.6	4.4	—23.6
29.0	28.1	22.8	19.2	16.6	8.9	10.4	— 4.1	4.2	—24.6
28.8	27.8	22.6	18.9	16.4	8.6	10.2	— 4.6		
28.6	27.5	22.4	18.6	16.2	8.2			4.0	—25.6
28.4	27.3	22.2	18.3			10.0	— 5.1	3.8	—26.6
28.2	27.0			16.0	7.8	9.8	— 5.6	3.6	—27.6
		22.0	18.0	15.8	7.4	9.6	— 6.1	3.4	—28.7
28.0	26.7	21.8	17.7	15.6	7.1	9.4	— 6.7	3.2	—29.9
27.8	26.4	21.6	17.4	15.4	6.7	9.2	— 7.2	3.0	—31.1
27.6	26.2	21.4	17.1	15.2	6.3	9.0	— 7.7	2.8	—32.4
27.4	25.9	21.2	16.7	15.0	5.9	8.8	— 8.3	2.6	—33.7
27.2	25.6	21.0	16.4	14.8	5.5	8.6	— 8.8	2.4	—35.2
27.0	25.3	20.8	16.1	14.6	5.1	8.4	— 9.4	2.2	—36.7
26.8	25.0	20.6	15.8	14.4	4.7	8.2	—10.0		
26.6	24.8	20.4	15.5	14.2	4.3			2.0	—38.4
26.4	24.5	20.2	15.1			8.0	—10.6	1.8	—40.2
26.2	24.2			14.0	3.9	7.8	—11.2	1.6	—42.2
		20.0	14.8	13.8	3.5	7.6	—11.8	1.4	—44.5
26.0	23.9	19.8	14.5	13.6	3.1	7.4	—12.4	1.2	—47.0
25.8	23.6	19.6	14.1	13.4	2.7	7.2	—13.0	1.0	—49.9

k = 33									
s	*l*	*s*	*l*	*s*	*l*	*s*	*l*	*s*	*l*
33.0	33.0								
32.8	32.7	26.4	24.1	20.0	14.4	13.6	2.5	7.2	—13.9
32.6	32.5	26.2	23.9	19.8	14.0	13.4	2.1		
32.4	32.2	26.0	23.6	19.6	13.7	13.2	1.7	7.0	—14.5
32.2	32.0	25.8	23.3	19.4	13.4			6.8	—15.2
32.0	31.7	25.6	23.0	19.2	13.0	13.0	1.3	6.6	—15.9
31.8	31.5	25.4	22.7			12.8	0.8	6.4	—16.6
31.6	31.2	25.2	22.4	19.0	12.7	12.6	0.4	6.2	—17.3
31.4	30.9			18.8	12.3	12.4	— 0.1	6.0	—18.0
31.2	30.7	25.0	22.1	18.6	12.0	12.2	— 0.5	5.8	—18.8
		24.8	21.8	18.4	11.7	12.0	— 1.0	5.6	—19.5
31.0	30.4	24.6	21.5	18.2	11.3	11.8	— 1.4	5.4	—20.3
30.8	30.2	24.4	21.2	18.0	11.0	11.6	— 1.9	5.2	—21.1
30.6	29.9	24.2	20.9	17.8	10.6	11.4	— 2.4		
30.4	29.6	24.0	20.7	17.6	10.2	11.2	— 2.8	5.0	—22.0
30.2	29.4	23.8	20.4	17.4	9.9			4.8	—22.8
30.0	29.1	23.6	20.1	17.2	9.5	11.0	— 3.3	4.6	—23.7
29.8	28.8	23.4	19.7			10.8	— 3.8	4.4	—24.7
29.6	28.6	23.2	19.4	17.0	9.2	10.6	— 4.3	4.2	—25.6
29.4	28.3			16.8	8.8	10.4	— 4.8	4.0	—26.6
29.2	28.0	23.0	19.1	16.6	8.4	10.2	— 5.3	3.8	—27.6
		22.8	18.8	16.4	8.1	10.0	— 5.8	3.6	—28.7
29.0	27.8	22.6	18.5	16.2	7.7	9.8	— 6.3	3.4	—29.8
28.8	27.5	22.4	18.2	16.0	7.3	9.6	— 6.8	3.2	—31.0
28.6	27.2	22.2	17.9	15.8	6.9	9.4	— 7.4		
28.4	26.9	22.0	17.6	15.6	6.5	9.2	— 7.9	3.0	—32.3
28.2	26.7	21.8	17.3	15.4	6.2			2.8	—33.6
28.0	26.4	21.6	17.0	15.2	5.8	9.0	— 8.5	2.6	—35.0
27.8	26.1	21.4	16.6			8.8	— 9.0	2.4	—36.4
27.6	25.8	21.2	16.3	15.0	5.4	8.6	— 9.6	2.2	—38.0
27.4	25.6			14.8	5.0	8.4	—10.2	2.0	—39.7
27.2	25.3	21.0	16.0	14.6	4.6	8.2	—10.8	1.8	—41.6
		20.8	15.7	14.4	4.2	8.0	—11.4	1.6	—43.7
27.0	25.0	20.6	15.4	14.2	3.8	7.8	—12.0	1.4	—46.0
26.8	24.7	20.4	15.0	14.0	3.4	7.6	—12.6	1.2	—48.6
26.6	24.4	20.2	14.7	13.8	3.0	7.4	—13.2	1.0	—51.6

$k = 34$

s	*l*	*s*	*l*	*s*	*l*	*s*	*l*	*s*	*l*
34.0	**34.0**								
33.8	**33.7**	27.2	24.9	20.6	14.9	14.0	2.8	7.4	—14.0
33.6	**33.5**	27.0	24.7	20.4	14.6	13.8	2.4	7.2	—14.7
33.4	**33.2**	26.8	24.4	20.2	14.3	13.6	2.0	7.0	—15.3
33.2	**33.0**	26.6	24.1			13.4	1.5	6.8	—16.0
33.0	**32.7**	26.4	23.8	20.0	13.9	13.2	1.1	6.6	—16.7
32.8	**32.5**	26.2	23.5	19.8	13.6	13.0	0.7	6.4	—17.4
32.6	**32.2**			19.6	13.3	12.8	0.2	6.2	—18.2
32.4	**32.0**	26.0	23.2	19.4	12.9	12.6	— 0.2		
32.2	**31.7**	25.8	22.9	19.2	12.6	12.4	— 0.7	6.0	—18.9
		25.6	22.6	19.0	12.2	12.2	— 1.1	5.8	—19.7
32.0	**31.4**	25.4	22.4	18.8	11.9			5.6	—20.5
31.8	**31.2**	25.2	22.1	18.6	11.5	12.0	— 1.6	5.4	—21.3
31.6	**30.9**	25.0	21.8	18.4	11.2	11.8	— 2.0	5.2	—22.1
31.4	**30.7**	24.8	21.5	18.2	10.8	11.6	— 2.5	5.0	—22.9
31.2	**30.4**	24.6	21.2			11.4	— 3.0	4.8	—23.8
31.0	**30.1**	24.4	20.9	18.0	10.5	11.2	— 3.5	4.6	—24.7
30.8	**29.9**	24.2	20.6	17.8	10.1	11.0	— 4.0	4.4	—25.7
30.6	**29.6**			17.6	9.8	10.8	— 4.4	4.2	—26.7
30.4	**29.3**	24.0	20.3	17.4	9.4	10.6	— 4.9		
30.2	**29.1**	23.8	20.0	17.2	9.0	10.4	— 5.5	4.0	—27.7
		23.6	19.7	17.0	8.7	10.2	— 6.0	3.8	—28.7
30.0	**28.8**	23.4	19.4	16.8	8.3			3.6	—29.8
29.8	**28.5**	23.2	19.1	16.6	7.9	10.0	— 6.5	3.4	—31.0
29.6	**28.3**	23.0	18.8	16.4	7.6	9.8	— 7.0	3.2	—32.2
29.4	**28.0**	22.8	18.4	16.2	7.2	9.6	— 7.5	3.0	—33.4
29.2	**27.7**	22.6	18.1			9.4	— 8.1	2.8	—34.8
29.0	**27.4**	22.4	17.8	16.0	6.8	9.2	— 8.6	2.6	—36.2
28.8	**27.2**	22.2	17.5	15.8	6.4	9.0	— 9.2	2.4	—37.7
28.6	**26.9**			15.6	6.0	8.8	— 9.8	2.2	—39.3
28.4	**26.6**	22.0	17.2	15.4	5.6	8.6	—10.3		
28.2	**26.3**	21.8	16.9	15.2	5.2	8.4	—10.9	2.0	—41.1
		21.6	16.6	15.0	4.8	8.2	—11.5	1.8	—43.0
28.0	**26.1**	21.4	16.2	14.8	4.4			1.6	—45.1
27.8	**25.8**	21.2	15.9	14.6	4.0	8.0	—12.1	1.4	—47.5
27.6	**25.5**	21.0	15.6	14.4	3.6	7.8	—12.7	1.2	—50.1
27.4	**25.2**	20.8	15.3	14.2	3.2	7.6	—13.4	1.0	—53.2

k = 35

s	*l*	*s*	*l*	*s*	*l*	*s*	*l*	*s*	*l*
35.0	35.0								
34.8	34.7	28.0	25.7	21.2	15.5	14.6	3.5	7.8	—13.5
34.6	34.5	27.8	25.5			14.4	3.1	7.6	—14.2
34.4	34.2	27.6	25.2	21.0	15.2	14.2	2.7	7.4	—14.8
34.2	34.0	27.4	24.9	20.8	14.8	14.0	2.2	7.2	—15.5
34.0	33.7	27.2	24.6	20.6	14.5	13.8	1.8		
33.8	33.5			20.4	14.2	13.6	1.4	7.0	—16.2
33.6	33.2	27.0	24.3	20.2	13.8	13.4	1.0	6.8	—16.9
33.4	33.0	26.8	24.0	20.0	13.5	13.2	0.5	6.6	—17.6
33.2	32.7	26.6	23.8	19.8	13.2			6.4	—18.3
		26.4	23.5	19.6	12.8	13.0	0.1	6.2	—19.0
33.0	32.5	26.2	23.2	19.4	12.5	12.8	— 0.4	6.0	—19.8
32.8	32.2	26.0	22.9	19.2	12.1	12.6	— 0.8	5.8	—20.6
32.6	31.9	25.8	22.6			12.4	— 1.3	5.6	—21.4
32.4	31.7	25.6	22.3	19.0	11.8	12.2	— 1.7	5.4	—22.2
32.2	31.4	25.4	22.0	18.8	11.4	12.0	— 2.2	5.2	—23.0
32.0	31.2	25.2	21.7	18.6	11.1	11.8	— 2.7		
31.8	30.9			18.4	10.7	11.6	— 3.1	5.0	—23.9
31.6	30.6	25.0	21.4	18.2	10.4	11.4	— 3.6	4.8	—24.8
31.4	30.4	24.8	21.1	18.0	10.0	11.2	— 4.1	4.6	—25.7
31.2	30.1	24.6	20.8	17.8	9.7			4.4	—26.7
		24.4	20.5	17.6	9.3	11.0	— 4.6	4.2	—27.7
31.0	29.8	24.2	20.2	17.4	8.9	10.8	— 5.1	4.0	—28.7
30.8	29.6	24.0	19.9	17.2	8.6	10.6	— 5.6	3.8	—29.8
30.6	29.3	23.8	19.6			10.4	— 6.1	3.6	—30.9
30.4	29.0	23.6	19.3	17.0	8.2	10.2	— 6.6	3.4	—32.1
30.2	28.8	23.4	19.0	16.8	7.8	10.0	— 7.2	3.2	—33.3
30.0	28.5	23.2	18.7	16.6	7.4	9.8	— 7.7		
29.8	28.2			16.4	7.1	9.6	— 8.2	3.0	—34.6
29.6	27.9	23.0	18.4	16.2	6.7	9.4	— 8.8	2.8	—36.0
29.4	27.7	22.8	18.1	16.0	6.3	9.2	— 9.4	2.6	—37.4
29.2	27.4	22.6	17.7	15.8	5.9			2.4	—39.0
		22.4	17.4	15.6	5.5	9.0	— 9.9	2.2	—40.6
29.0	27.1	22.2	17.1	15.4	5.1	8.8	—10.5	2.0	—42.4
28.8	26.9	22.0	16.8	15.2	4.7	8.6	—11.1	1.8	—44.4
28.6	26.6	21.8	16.5			8.4	—11.7	1.6	—46.6
28.4	26.3	21.6	16.1	15.0	4.3	8.2	—12.3	1.4	—49.0
28.2	26.0	21.4	15.8	14.8	3.9	8.0	—12.9	1.2	—51.7
								1.0	—54.9

$k = 36$									
s	*l*	*s*	*l*	*s*	*l*	*s*	*l*	*s*	*l*
36.0	36.0								
35.8	35.7	28.8	26.5	22.0	16.4	14.8	3.4	8.0	—13.7
35.6	35.5	28.6	26.3	21.8	16.1	14.6	3.0	7.8	—14.3
35.4	35.2	28.4	26.0	21.6	15.7	14.4	2.5	7.6	—15.0
35.2	35.0	28.2	25.7	21.4	15.4	14.2	2.1	7.4	—15.6
35.0	34.7			21.2	15.1			7.2	—16.3
34.8	34.5	28.0	25.4	21.0	14.8	14.0	1.7	7.0	—17.0
34.6	34.2	27.8	25.1	20.8	14.4	13.8	1.3	6.8	—17.7
34.4	34.0	27.6	24.8	20.6	14.1	13.6	0.8	6.6	—18.4
34.2	33.7	27.4	24.6	20.4	13.8	13.4	0.4	6.4	—19.2
		27.2	24.3	20.2	13.4	13.2	— 0.1	6.2	—19.9
34.0	33.5	27.0	24.0			13.0	— 0.5		
33.8	33.2	26.8	23.7	20.0	13.1	12.8	— 1.0	6.0	—20.7
33.6	33.0	26.6	23.4	19.8	12.7	12.6	— 1.4	5.8	—21.5
33.4	32.7	26.4	23.1	19.6	12.4	12.4	— 1.9	5.6	—22.3
33.2	32.4	26.2	22.8	19.4	12.0	12.2	— 2.3	5.4	—23.1
33.0	32.2			19.2	11.7			5.2	—24.0
32.8	31.9	26.0	22.5	19.0	11.3	12.0	— 2.8	5.0	—24.9
32.6	31.7	25.8	22.2	18.8	11.0	11.8	— 3.3	4.8	—25.8
32.4	31.4	25.6	21.9	18.6	10.6	11.6	— 3.8	4.6	—26.7
32.2	31.1	25.4	21.6	18.4	10.3	11.4	— 4.3	4.4	—27.7
		25.2	21.4	18.2	9.9	11.2	— 4.8	4.2	—28.7
32.0	30.9	25.0	21.1			11.0	— 5.3		
31.8	30.6	24.8	20.8	18.0	9.6	10.8	— 5.8	4.0	—29.8
31.6	30.3	24.6	20.4	17.8	9.2	10.6	— 6.3	3.8	—30.9
31.4	30.1	24.4	20.1	17.6	8.8	10.4	— 6.8	3.6	—32.0
31.2	29.8	24.2	19.8	17.4	8.4	10.2	— 7.3	3.4	—33.2
31.0	29.5			17.2	8.1			3.2	—34.5
30.8	29.3	24.0	19.5	17.0	7.7	10.0	— 7.9	3.0	—35.8
30.6	29.0	23.8	19.2	16.8	7.3	9.8	— 8.4	2.8	—37.2
30.4	28.7	23.6	18.9	16.6	6.9	9.6	— 8.9	2.6	—38.7
30.2	28.5	23.4	18.6	16.4	6.6	9.4	— 9.5	2.4	—40.3
		23.2	18.3	16.2	6.2	9.2	—10.1	2.2	—42.0
30.0	28.2	23.0	18.0			9.0	—10.6		
29.8	27.9	22.8	17.7	16.0	5.8	8.8	—11.2	2.0	—43.8
29.6	27.6	22.6	17.4	15.8	5.4	8.6	—11.8	1.8	—45.8
29.4	27.4	22.4	17.0	15.6	5.0	8.4	—12.4	1.6	—48.0
29.2	27.1	22.2	16.7	15.4	4.6	8.2	—13.0	1.4	—50.5
29.0	26.8			15.2	4.2			1.2	—53.2
				15.0	3.8			1.0	—56.5

$k = 37$

s	*l*	*s*	*l*	*s*	*l*	*s*	*l*	*s*	*l*
37.0	37.0								
36.8	36.8	29.6	27.3	22.4	16.6	15.0	3.2	8.0	—14.4
36.6	36.5	29.4	27.1	22.2	16.3	14.8	2.8	7.8	—15.1
36.4	36.3	29.2	26.8	22.0	16.0	14.6	2.4	7.6	—15.8
36.2	36.0			21.8	15.7	14.4	2.0	7.4	—16.4
36.0	35.8	29.0	26.5	21.6	15.3	14.2	1.6	7.2	—17.1
35.8	35.5	28.8	26.2	21.4	15.0	14.0	1.1		
35.6	35.2	28.6	25.9	21.2	14.7	13.8	0.7	7.0	—17.8
35.4	35.0	28.4	25.7			13.6	0.3	6.8	—18.5
35.2	34.7	28.2	25.4	21.0	14.3	13.4	— 0.2	6.6	—19.3
		28.0	25.1	20.8	14.0	13.2	— 0.6	6.4	—20.0
35.0	34.5	27.8	24.8	20.6	13.7			6.2	—20.8
34.8	34.2	27.6	24.5	20.4	13.3	13.0	— 1.1	6.0	—21.6
34.6	34.0	27.4	24.2	20.2	13.0	12.8	— 1.6	5.8	—22.4
34.4	33.7	27.2	23.9	20.0	12.6	12.6	— 2.0	5.6	—23.2
34.2	33.5			19.8	12.3	12.4	— 2.5	5.4	—24.1
34.0	33.2	27.0	23.7	19.6	11.9	12.2	— 3.0	5.2	—24.9
33.8	32.9	26.8	23.4	19.4	11.6	12.0	— 3.4		
33.6	32.7	26.6	23.1	19.2	11.2	11.8	— 3.9	5.0	—25.8
33.4	32.4	26.4	22.8			11.6	— 4.4	4.8	—26.8
33.2	32.2	26.2	22.5	19.0	10.9	11.4	— 4.9	4.6	—27.7
		26.0	22.2	18.8	10.5	11.2	— 5.4	4.4	—28.7
33.0	31.9	25.8	21.9	18.6	10.2			4.2	—29.8
32.8	31.6	25.6	21.6	18.4	9.8	11.0	— 5.9	4.0	—30.8
32.6	31.4	25.4	21.3	18.2	9.4	10.8	— 6.4	3.8	—32.0
32.4	31.1	25.2	21.0	18.0	9.1	10.6	— 6.9	3.6	—33.1
32.2	30.9	25.0	20.7	17.8	8.7	10.4	— 7.5	3.4	—34.3
32.0	30.6			17.6	8.3	10.2	— 8.0	3.2	—35.6
31.8	30.3	24.8	20.4	17.4	8.0	10.0	— 8.5		
31.6	30.1	24.6	20.1	17.2	7.6	9.8	— 9.1	3.0	—37.0
31.4	29.8	24.4	19.8			9.6	— 9.7	2.8	—38.4
31.2	29.5	24.2	19.5	17.0	7.2	9.4	—10.2	2.6	—39.9
		24.0	19.2	16.8	6.8	9.2	—10.8	2.4	—41.5
31.0	29.2	23.8	18.9	16.6	6.4			2.2	—43.3
30.8	29.0	23.6	18.5	16.4	6.1	9.0	—11.4	2.0	—45.1
30.6	28.7	23.4	18.2	16.2	5.7	8.8	—12.0	1.8	—47.2
30.4	28.4	23.2	17.9	16.0	5.3	8.6	—12.6	1.6	—49.4
30.2	28.2			15.8	4.9	8.4	—13.2	1.4	—52.0
30.0	27.9	23.0	17.6	15.6	4.5	8.2	—13.8	1.2	—54.8
29.8	27.6	22.8	17.3	15.4	4.1			1.0	—58.1
		22.6	17.0	15.2	3.7				

$k = 38$									
s	l	s	l	s	l	s	l	s	l
38.0	38.0								
37.8	37.8	30.4	28.1	23.0	17.2	15.6	3.9	8.0	−15.2
37.6	37.5	30.2	27.9	22.8	16.9	15.4	3.5	7.8	−15.9
37.4	37.3			22.6	16.6	15.2	3.1	7.6	−16.6
37.2	37.0	30.0	27.6	22.4	16.2	15.0	2.7	7.4	−17.2
37.0	36.8	29.8	27.3	22.2	15.9	14.8	2.3	7.2	−17.9
36.8	36.5	29.6	27.0			14.6	1.9	7.0	−18.6
36.6	36.3	29.4	26.4	22.0	15.6	14.4	1.4	6.8	−19.4
36.4	36.0	29.2	26.5	21.8	15.3	14.2	1.0	6.6	−20.1
36.2	35.8	29.0	26.2	21.6	14.9			6.4	−20.9
		28.8	25.9	21.4	14.6	14.0	0.6	6.2	−21.7
36.0	35.5	28.6	25.6	21.2	14.3	13.8	0.1		
35.8	35.2	28.4	25.3	21.0	13.9	13.6	−0.3	6.0	−22.5
35.6	35.0	28.2	25.1	20.8	13.6	13.4	−0.8	5.8	−23.3
35.4	34.7			20.6	13.2	13.2	−1.2	5.6	−24.1
35.2	34.5	28.0	24.8	20.4	12.9	13.0	−1.7	5.4	−25.0
35.0	34.2	27.8	24.5	20.2	12.6	12.8	−2.1	5.2	−25.9
34.8	34.0	27.6	24.2			12.6	−2.6	5.0	−26.8
34.6	33.7	27.4	23.9	20.0	12.2	12.4	−3.1	4.8	−27.7
34.4	33.5	27.2	23.6	19.8	11.9	12.2	−3.6	4.6	−28.7
34.2	33.2	27.0	23.3	19.6	11.5			4.4	−29.7
		26.8	23.0	19.4	11.2	12.0	−4.0	4.2	−30.8
34.0	32.9	26.6	22.7	19.2	10.8	11.8	−4.5		
33.8	32.7	26.4	22.4	19.0	10.4	11.6	−5.0	4.0	−31.9
33.6	32.4	26.2	22.1	18.8	10.1	11.4	−5.5	3.8	−33.0
33.4	32.2			18.6	9.7	11.2	−6.0	3.6	−34.2
33.2	31.9	26.0	21.8	18.4	9.4	11.0	−6.6	3.4	−35.5
33.0	31.6	25.8	21.5	18.2	9.0	10.8	−7.1	3.2	−36.8
32.8	31.4	25.6	21.2			10.6	−7.6	3.0	−38.1
32.6	31.1	25.4	20.9	18.0	8.6	10.4	−8.1	2.8	−39.6
32.4	30.8	25.2	20.6	17.8	8.2	10.2	−8.7	2.6	−41.1
32.2	30.6	25.0	20.3	17.6	7.9			2.4	−42.8
		24.8	20.0	17.4	7.5	10.0	−9.2	2.2	−44.6
32.0	30.3	24.6	19.7	17.2	7.1	9.8	−9.8		
31.8	30.0	24.4	19.4	17.0	6.7	9.6	−10.4	2.0	−46.5
31.6	29.8	24.2	19.1	16.8	6.3	9.4	−10.9	1.8	−48.6
31.4	29.5			16.6	5.9	9.2	−11.5	1.6	−50.9
31.2	29.2	24.0	18.8	16.4	5.5	9.0	−12.1	1.4	−53.5
31.0	29.0	23.8	18.5	16.2	5.2	8.8	−12.7	1.2	−56.4
30.8	28.7	23.6	18.2			8.6	−13.3	1.0	−59.8
30.6	28.4	23.4	17.8	16.0	4.8	8.4	−13.9		
		23.2	17.5	15.8	4.4	8.2	−14.6		

$k = 39$

s	*l*	*s*	*l*	*s*	*l*	*s*	*l*	*s*	*l*
39.0	39.0								
38.8	38.8	31.0	28.7	23.6	17.8	16.0	4.2	8.4	—14.7
38.6	38.5	30.8	28.4	23.4	17.5	15.8	3.8	8.2	—15.3
38.4	38.3	30.6	28.1	23.2	17.1	15.6	3.4	8.0	—16.0
38.2	38.0	30.4	27.8			15.4	3.0	7.8	—16.7
38.0	37.8	30.2	27.6	23.0	16.8	15.2	2.6	7.6	—17.3
37.8	37.5	30.0	27.3	22.8	16.5			7.4	—18.0
37.6	37.3	29.8	27.0	22.6	16.2	15.0	2.2	7.2	—18.7
37.4	37.0	29.6	26.7	22.4	15.8	14.8	1.7		
37.2	36.8	29.4	26.4	22.2	15.5	14.6	1.3	7.0	—19.5
		29.2	26.2	22.0	15.2	14.4	0.9	6.8	—20.2
37.0	36.5			21.8	14.9	14.2	0.4	6.6	—21.0
36.8	36.3	29.0	25.9	21.6	14.5	14.0	0.0	6.4	—21.7
36.6	36.0	28.8	25.6	21.4	14.2	13.8	— 0.4	6.2	—22.5
36.4	35.8	28.6	25.3	21.2	13.8	13.6	— 0,9	6.0	—23.3
36.2	35.5	28.4	25.0			13.4	— 1.3	5.8	—24.2
36.0	35.3	28.2	24.7	21.0	13.5	13.2	— 1.8	5.6	—25.0
35.8	35.0	28.0	24.4	20.8	13.2			5.4	—25.9
35.6	34.7	27.8	24.2	20.6	12.8	13.0	— 2.3	5.2	—26.8
35.4	34.5	27.6	23.9	20.4	12.5	12.8	— 2.7		
35.2	34.2	27.4	23.6	20.2	12.1	12.6	— 3.2	5.0	—27.8
		27.2	23.3	20.0	11.8	12.4	— 3.7	4.8	—28.7
35.0	34.0			19.8	11.4	12.2	— 4.2	4.6	—29.7
34.8	33.7	27.0	23.0	19.6	11.1	12.0	— 4.7	4.4	—30.8
34.6	33.5	26.8	22.7	19.4	10.7	11.8	— 5.2	4.2	—31.8
34.4	33.2	26.6	22.4	19.2	10.4	11.6	— 5.7	4.0	—32.9
34.2	32.9	26.4	22.1			11.4	— 6.2	3.8	—34.1
34.0	32.7	26.2	21.8	19.0	10.0	11.2	— 6.7	3.6	—35.3
33.8	32.4	26.0	21.5	18.8	9.6			3.4	—36.6
33.6	32.1	25.8	21.2	18.6	9.3	11.0	— 7.2	3.2	—37.9
33.4	31.9	25.6	20.9	18.4	8.9	10.8	— 7.7		
33.2	31.6	25.4	20.6	18.2	8.5	10.6	— 8.3	3.0	—39.3
		25.2	20.3	18.0	8.1	10.4	— 8.8	2.8	—40.8
33.0	31.4			17.8	7.8	10.2	— 9.4	2.6	—42.4
32.8	31.1	25.0	20.0	17.6	7.4	10.0	— 9.9	2.4	—44.1
32.6	30.8	24.8	19.7	17.4	7.0	9.8	—10.5	2.2	—45.9
32.4	30.6	24.6	19.4	17.2	6.6	9.6	—11.1	2.0	—47.8
32.2	30.3	24.4	19.0			9.4	—11.6	1.8	—50.0
32.0	30.0	24.2	18.7	17.0	6.2	9.2	—12.2	1.6	—52.3
31.8	29.8	24.0	18.4	16.8	5.8			1.4	—54.9
31.6	29.5	23.8	18.1	16.6	5.4	9.0	—12.8	1.2	—57.9
31.4	29.2			16.4	5.0	8.8	—13.4	1.0	—61.4
31.2	28.9			16.2	4.6	8.6	—14.1		

				$k = 40$					
s	*l*	*s*	*l*	*s*	*l*	*s*	*l*	*s*	*l*
40.0	40.0								
39.8	39.8	32.0	29.7	24.0	18.0	16.0	3.7	8.0	−16.8
39.6	39.5	31.8	29.5	23.8	17.7	15.8	3.3	7.8	−17.4
39.4	39.3	31.6	29.2	23.6	17.4	15.6	2.9	7.6	−18.1
39.2	39.0	31.4	28.9	23.4	17.1	15.4	2.5	7.4	−18.8
39.0	38.8	31.2	28.6	23.2	16.8	15.2	2.1	7.2	−19.6
38.8	38.5	31.0	28.4	23.0	16.4	15.0	1.6	7.0	−20.3
38.6	38.3	30.8	28.1	22.8	16.1	14.8	1.2	6.8	−21.0
38.4	38.0	30.6	27.8	22.6	15.8	14.6	0.8	6.6	−21.8
38.2	37.8	30.4	27.5	22.4	15.5	14.4	0.3	6.4	−22.6
		30.2	27.3	22.2	15.1	14.2	−0.1	6.2	−23.4
38.0	37.5								
37.8	37.3	30.0	27.0	22.0	14.8	14.0	−0.6	6.0	−24.2
37.6	37.0	29.8	26.7	21.8	14.5	13.8	−1.0	5.8	−25.1
37.4	36.8	29.6	26.4	21.6	14.1	13.6	−1.5	5.6	−26.0
37.2	36.5	29.4	26.1	21.4	13.8	13.4	−1.9	5.4	−26.9
37.0	36.3	29.2	25.8	21.2	13.4	13.2	−2.4	5.2	−27.8
36.8	36.0	29.0	25.6	21.0	13.1	13.0	−2.9	5.0	−28.7
36.6	35.8	28.8	25.3	20.8	12.7	12.8	−3.3	4.8	−29.7
36.4	35.5	28.6	25.0	20.6	12.4	12.6	−3.8	4.6	−30.7
36.2	35.3	28.4	24.7	20.4	12.0	12.4	−4.3	4.4	−31.8
		28.2	24.4	20.2	11.7	12.2	−4.8	4.2	−32.9
36.0	35.0								
35.8	34.7	28.0	24.1	20.0	11.3	12.0	−5.3	4.0	−34.0
35.6	34.5	27.8	23.8	19.8	11.0	11.8	−5.8	3.8	−35.2
35.4	34.2	27.6	23.5	19.6	10.6	11.6	−6.3	3.6	−36.4
35.2	34.0	27.4	23.2	19.4	10.3	11.4	−6.8	3.4	−37.7
35.0	33.7	27.2	22.9	19.2	9.9	11.2	−7.3	3.2	−39.1
34.8	33.5	27.0	22.7	19.0	9.5	11.0	−7.9	3.0	−40.5
34.6	33.2	26.8	22.4	18.8	9.2	10.8	−8.4	2.8	−42.0
34.4	32.9	26.6	22.1	18.6	8.8	10.6	−8.9	2.6	−43.6
34.2	32.7	26.4	21.8	18.4	8.4	10.4	−9.5	2.4	−45.3
		26.2	21.5	18.2	8.1	10.2	−10.0	2.2	−47.2
34.0	32.4								
33.8	32.1	26.0	21.2	18.0	7.7	10.0	−10.6	2.0	−49.2
33.6	31.9	25.8	20.8	17.8	7.3	9.8	−11.2	1.8	−51.4
33.4	31.6	25.6	20.5	17.6	6.9	9.6	−11.8	1.6	−53.8
33.2	31.3	25.4	20.2	17.4	6.5	9.4	−12.4	1.4	−56.4
33.0	31.1	25.2	19.9	17.2	6.1	9.2	−13.0	1.2	−59.5
32.8	30.8	25.0	19.6	17.0	5.7	9.0	−13.6	1.0	−63.1
32.6	30.5	24.8	19.3	16.8	5.3	8.8	−14.2		
32.4	30.3	24.6	19.0	16.6	4.9	8.6	−14.8		
32.2	30.0	24.4	18.7	16.4	4.5	8.4	−15.5		
		24.2	18.4	16.2	4.1	8.2	−16.1		

$k = 41$									
s	*l*	*s*	*l*	*s*	*l*	*s*	*l*	*s*	*l*
41.0	41.0								
40.8	40.8	33.8	31.9	27.0	22.3	20.0	10.9	13.0	— 3.4
40.6	40.5	33.6	31.6	26.8	22.0	19.8	10.5	12.8	— 3.9
40.4	40.3	33.4	31.3	26.6	21.7	19.6	10.2	12.6	— 4.4
40.2	40.0	33.2	31.1	26.4	21.4	19.4	9.8	12.4	— 4.9
40.0	39.8			26.2	21.1	19.2	9.5	12.2	— 5.4
39.8	39.5	33.0	30.8	26.0	20.8			12.0	— 5.9
39.6	39.3	32.8	30.5	25.8	20.5	19.0	9.1	11.8	— 6.4
39.4	39.0	32.6	30.3	25.6	20.2	18.8	8.7	11.6	— 6.9
39.2	38.8	32.4	30.0	25.4	19.9	18.6	8.3	11.4	— 7.4
		32.2	29.7	25.2	19.6	18.4	8.0	11.2	— 8.0
39.0	38.5	32.0	29.5			18.2	7.6		
38.8	38.3	31.8	29.2	25.0	19.3	18.0	7.2	11.0	— 8.5
38.6	38.0	31.6	28.9	24.8	18.9	17.8	6.8	10.8	— 9.0
38.4	37.8	31.4	28.6	24.6	18.6	17.6	6.4	10.6	— 9.6
38.2	37.5	31.2	28.4	24.4	18.3	17.4	6.0	10.4	—10.2
38.0	37.3			24.2	18.0	17.2	5.6	10.2	—10.7
37.8	37.0	31.0	28.1	24.0	17.7			10.0	—11.3
37.6	36.8	30.8	27.8	23.8	17.4	17.0	5.3	9.8	—11.9
37.4	36.5	30.6	27.5	23.6	17.0	16.8	4.9	9.6	—12.5
37.2	36.3	30.4	27.2	23.4	16.7	16.6	4.4	9.4	—13.1
		30.2	27.0	23.2	16.4	16.4	4.0	9.2	—13.7
37.0	36.0	30.0	26.7			16.2	3.6		
36.8	35.8	29.8	26.4	23.0	16.1	16.0	3.2	9.0	—14.3
36.6	35.5	29.6	26.1	22.8	15.7	15.8	2.8	8.8	—14.9
36.4	35.3	29.4	25.8	22.6	15.4	15.6	2.4	8.6	—15.6
36.2	35.0	29.2	25.5	22.4	15.1	15.4	2.0	8.4	—16.2
36.0	34.8			22.2	14.7	15.2	1.5	8.2	—16.9
35.8	34.5	29.0	25.3	22.0	14.4			8.0	—17.5
35.6	34.2	28.8	25.0	21.8	14.0	15.0	1.1	7.8	—18.2
35.4	34.0	28.6	24.7	21.6	13.7	14.8	0.7	7.6	—18.9
35.2	33.7	28.4	24.4	21.4	13.4	14.6	0.2	7.4	—19.6
		28.2	24.1	21.2	13.0	14.4	—0.2	7.2	—20.4
35.0	33.5	28.0	23.8			14.2	—0.7		
34.8	33.2	27.8	23.5	21.0	12.7	14.0	—1.1	7.0	—21.1
34.6	32.9	27.6	23.2	20.8	12.3	13.8	—1.6	6.8	—21.9
34.4	32.7	27.4	22.9	20.6	12.0	13.6	—2.0	6.6	—22.7
34.2	32.4	27.2	22.6	20.4	11.6	13.4	—2.5	6.4	—23.5
34.0	32.1			20.2	11.3	13.2	—3.0	6.2	—24.3

$k = 41$									
s	l	s	l	s	l	s	l	s	l
6.0	—25.1	5.0	—29.7	4.0	—35.1	3.0	—41.7	2.0	—50.5
5.8	—26.0	4.8	—30.7	3.8	—36.3	2.8	—43.2	1.8	—52.7
5.6	—26.9	4.6	—31.7	3.6	—37.5	2.6	—44.9	1.6	—55.2
5.4	—27.8	4.4	—32.8	3.4	—38.8	2.4	—46.6	1.4	—57.9
5.2	—28.7	4.2	—33.9	3.2	—40.2	2.2	—48.5	1.2	—61.1
								1.0	—64.7

$k = 42$

s	*l*	*s*	*l*	*s*	*l*	*s*	*l*	*s*	*l*
42.0	42.0								
41.8	41.8	34.8	32.9	28.0	23.5	21.0	12.3	14.0	— 1.7
41.6	41.5	34.6	32.7	27.8	23.2	20.8	11.9	13.8	— 2.1
41.4	41.3	34.4	32.4	27.6	22.9	20.6	11.6	13.6	— 2.6
41.2	41.0	34.2	32.1	27.4	22.6	20.4	11.2	13.4	— 3.1
41.0	40.8			27.2	22.3	20.2	10.8	13.2	— 3.5
40.8	40.5	34.0	31.9	27.0	22.0			13.0	— 4.0
40.6	40.3	33.8	31.6	26.8	21.7	20.0	10.5	12.8	— 4.5
40.4	40.1	35.6	31.3	26.6	21.4	19.8	10.1	12.6	— 5.0
40.2	39.8	33.4	31.1	26.4	21.1	19.6	9.7	12.4	— 5.5
		33.2	30.8	26.2	20.8	19.4	9.4	12.2	— 6.0
40.0	39.6	33.0	30.5			19.2	9.0		
39.8	39.3	32.8	30.3	26.0	20.5	19.0	8.6	12.0	— 6.5
39.6	39.1	32.6	30.0	25.8	20.1	18.8	8.3	11.8	— 7.0
39.4	38.8	32.4	29.7	25.6	19.8	18.6	7.9	11.6	— 7.6
39.2	38.6	32.2	29.4	25.4	19.5	18.4	7.5	11.4	— 8.1
39.0	38.3			25.2	19.2	18.2	7.1	11.2	— 8.6
38.8	38.1	32.0	29.2	25.0	18.9			11.0	— 9.2
38.6	37.8	31.8	28.9	24.8	18.6	18.0	6.7	10.8	— 9.7
38.4	37.6	31.6	28.6	24.6	18.3	17.8	6.3	10.6	—10.3
38.2	37.3	31.4	28.3	24.4	17.9	17.6	6.0	10.4	—10.8
		31.2	28.1	24.2	17.6	17.4	5.6	10.2	—11.4
38.0	37.1	31.0	27.8			17.2	5.2		
37.8	36.8	30.8	27.5	24.0	17.3	17.0	4.8	10.0	—12.0
37.6	36.6	30.6	27.2	23.8	17.0	16.8	4.4	9.8	—12.6
37.4	36.3	30.4	26.9	23.6	16.7	16.6	3.9	9.6	—13.2
37.2	36.0	30.2	26.7	23.4	16.3	16.4	3.5	9.4	—13.8
37.0	35.8			23.2	16.0	16.2	3.1	9.2	—14.4
36.8	35.5	30.0	26.4	23.0	15.7			9.0	—15.0
36.6	35.3	29.8	26.1	22.8	15.3	16.0	2.7	8.8	—15.7
36.4	35.0	29.6	25.8	22.6	15.0	15.8	2.3	8.6	—16.3
36.2	34.8	29.4	25.5	22.4	14.7	15.6	1.9	8.4	—17.0
		29.2	25.2	22.2	14.3	15.4	1.4	8.2	—17.6
36.0	34.5	29.0	24.9			15.2	1.0		
35.8	34.2	28.8	24.6	22.0	14.0	15.0	0.6	8.0	—18.3
35.6	34.0	28.6	24.4	21.8	13.6	14.8	0.1	7.8	—19.0
35.4	33.7	28.4	24.1	21.6	13.3	14.6	—0.3	7.6	—19.7
35.2	33.5	28.2	23.8	21.4	13.0	14.4	—0.8	7.4	—20.5
35.0	33.2			21.2	12.6	14.2	—1.2	7.2	—21.2

$k = 42$									
s	*l*	*s*	*l*	*s*	*l*	*s*	*l*	*s*	*l*
7.0	—22.0	5.8	—26.9	4.4	—33.8	3.2	—41.4	2.0	—51.9
6.8	—22.7	5.6	—27.8	4.2	—34.9	3.0	—42.8	1.8	—54.1
6.6	—23.5	5.4	—28.7			2.8	—44.4	1.6	—56.6
6.4	—24.3	5.2	—29.7	4.0	—36.1	2.6	—46.1	1.4	—59.4
6.2	—25.2	5.0	—30.7	3.8	—37.3	2.4	—47.9	1.2	—62.6
		4.8	—31.7	3.6	—38.6	2.2	—49.8	1.0	—66.4
6.0	—26.0	4.6	—32.7	3.4	—40.0				

				$k = 43$					
s	*l*	*s*	*l*	*s*	*l*	*s*	*l*	*s*	*l*
43.0	43.0								
42.8	42.8	36.2	34.5	29.6	25.5	23.0	15.3	16.4	3.0
42.6	42.5	36.0	34.3	29.4	25.2	22.8	14.9	16.2	2.6
42.4	42.3	35.8	34.0	29.2	24.9	22.6	14.6	16.0	2.2
42.2	42.0	35.6	33.7			22.4	14.3	15.8	1.8
42.0	41.8	35.4	33.5	29.0	24.6	22.2	13.9	15.6	1.3
41.8	41.5	35.2	33.2	28.8	24.3	22.0	13.6	15.4	0.9
41.6	41.3			28.6	24.0	21.8	13.2	15.2	0.5
41.4	41.1	35.0	32.9	28.4	23.7	21.6	12.9		
41.2	40.8	34.8	32.7	28.2	23.5	21.4	12.5	15.0	0.0
		34.6	32.4	28.0	23.2	21.2	12.2	14.8	−0.4
41.0	40.6	34.4	32.1	27.8	22.9			14.6	−0.9
40.8	40.3	34.2	31.9	27.6	22.6	21.0	11.8	14.4	−1.3
40.6	40.1	34.0	31.6	27.4	22.3	20.8	11.5	14.2	−1.8
40.4	39.8	33.8	31.3	27.2	22.0	20.6	11.1	14.0	−2.2
40.2	39.6	33.6	31.1			20.4	10.8	13.8	−2.7
40.0	39.3	33.4	30.8	27.0	21.6	20.2	10.4	13.6	−3.2
39.8	39.1	33.2	30.5	26.8	21.3	20.0	10.0	13.4	−3.6
39.6	38.8			26.6	21.0	19.8	9.7	13.2	−4.1
39.4	38.6	33.0	30.3	26.4	20.7	19.6	9.3		
39.2	38.3	32.8	30.0	26.2	20.4	19.4	8.9	13.0	−4.6
		32.6	29.7	26.0	20.1	19.2	8.6	12.8	−5.1
39.0	38.1	32.4	29.4	25.8	19.8			12.6	−5.6
38.8	37.8	32.2	29.2	25.6	19.5	19.0	8.2	12.4	−6.1
38.6	37.6	32.0	28.9	25.4	19.2	18.8	7.8	12.2	−6.6
38.4	37.3	31.8	28.6	25.2	18.9	18.6	7.4	12.0	−7.1
38.2	37.1	31.6	28.3			18.4	7.0	11.8	−7.7
38.0	36.8	31.4	28.1	25.0	18.5	18.2	6.7	11.6	−8.2
37.8	36.6	31.2	27.8	24.8	18.2	18.0	6.3	11.4	−8.7
37.6	36.3			24.6	17.9	17.8	5.9	11.2	−9.3
37.4	36.1	31.0	27.5	24.4	17.6	17.6	5.5		
37.2	35.8	30.8	27.2	24.2	17.3	17.4	5.1	11.0	−9.8
		30.6	26.9	24.0	16.9	17.2	4.7	10.8	−10.4
37.0	35.6	30.4	26.6	23.8	16.6			10.6	−10.9
36.8	35.3	30.2	26.4	23.6	16.3	17.0	4.3	10.4	−11.5
36.6	35.0	30.0	26.1	23.4	15.9	16.8	3.9	10.2	−12.1
36.4	34.8	29.8	25.8	23.2	15.6	16.6	3.5	10.0	−12.7

$k = 43$									
s	*l*	*s*	*l*	*s*	*l*	*s*	*l*	*s*	*l*
9.8	—13.3	8.0	—19.1	6.2	—26.0	4.4	—34.8	2.6	—47.3
9.6	—13.9	7.8	—19.8	6.0	—26.9	4.2	—36.0	2.4	—49.1
9.4	—14.5	7.6	—20.5	5.8	—27.8	4.0	—37.2	2.2	—51.1
9.2	—15.1	7.4	—21.3	5.6	—28.7	3.8	—38.4	2.0	—53.2
		7.2	—22.0	5.4	—29.6	3.6	—39.7	1.8	—55.5
9.0	—15.7			5.2	—30.6	3.4	—41.1	1.6	—58.1
8.8	—16.4	7.0	—22.8			3.2	—42.5	1.4	—60.9
8.6	—17.0	6.8	—23.6	5.0	—31.6			1.2	—64.2
8.4	—17.7	6.6	—24.4	4.8	—32.6	3.0	—44.0	1.0	—68.0
8.2	—18.4	6.4	—25.2	4.6	—33.7	2.8	—45.6		

k = 44									
s	*l*	*s*	*l*	*s*	*l*	*s*	*l*	*s*	*l*
44.0	44.0								
43.8	43.8	37.2	35.6	30.6	26.6	24.0	16.6	17.4	4.6
43.6	43.5	37.0	35.3	30.4	26.3	23.8	16.2	17.2	4.2
43.4	43.3	36.8	35.1	30.2	26.1	23.6	15.9	17.0	3.8
43.2	43.0	36.6	34.8			23.4	15.6	16.8	3.4
43.0	42.8	36.4	34.5	30.0	25.8	23.2	15.2	16.6	3.0
42.8	42.6	36.2	34.3	29.8	25.5	23.0	14.9	16.4	2.5
42.6	42.3			29.6	25.2	22.8	14.6	16.2	2.1
42.4	42.1	36.0	34.0	29.4	24.9	22.6	14.2		
42.2	41.8	35.8	33.7	29.2	24.6	22.4	13.9	16.0	1.7
		35.6	33.5	29.0	24.3	22.2	13.5	15.8	1.3
42.0	41.6	35.4	33.2	28.8	24.0			15.6	0.8
41.8	41.3	35.2	33.0	28.6	23.7	22.0	13.2	15.4	0.4
41.6	41.1	35.0	32.7	28.4	23.4	21.8	12.8	15.2	− 0.1
41.4	40.9	34.8	32.4	28.2	23.1	21.6	12.5	15.0	− 0.5
41.2	40.6	34.6	32.2			21.4	12.1	14.8	− 1.0
41.0	40.4	34.4	31.9	28.0	22.8	21.2	11.8	14.6	− 1.4
40.8	40.1	34.2	31.6	27.8	22.5	21.0	11.4	14.4	− 1.9
40.6	39.9			27.6	22.2	20.8	11.1	14.2	− 2.3
40.4	39.6	34.0	31.3	27.4	21.9	20.6	10.7		
40.2	39.4	33.8	31.1	27.2	21.6	20.4	10.3	14.0	− 2.8
		33.6	30.8	27.0	21.3	20.2	10.0	13.8	− 3.3
40.0	39.1	33.4	30.5	26.8	21.0			13.6	− 3.7
39.8	38.9	33.2	30.3	26.6	20.7	20.0	9.6	13.4	− 4.2
39.6	38.6	33.0	30.0	26.4	20.4	19.8	9.2	13.2	− 4.7
39.4	38.4	32.8	29.7	26.2	20.1	19.6	8.9	13.0	− 5.2
39.2	38.1	32.6	29.4			19.4	8.5	12.8	− 5.7
39.0	37.9	32.4	29.2	26.0	19.8	19.2	8.1	12.6	− 6.2
38.8	37.6	32.2	28.9	25.8	19.5	19.0	7.7	12.4	− 6.7
38.6	37.4			25.6	19.1	18.8	7.4	12.2	− 7.2
38.4	37.1	32.0	28.6	25.4	18.8	18.6	7.0		
38.2	36.8	31.8	28.3	25.2	18.5	18.4	6.6	12.0	− 7.8
		31.6	28.0	25.0	18.2	18.2	6.2	11.8	− 8.3
38.0	36.6	31.4	27.8	24.8	17.9			11.6	− 8.8
37.8	36.3	31.2	27.5	24.6	17.5	18.0	5.8	11.4	− 9.4
37.6	36.1	31.0	27.2	24.4	17.2	17.8	5.4	11.2	− 9.9
37.4	35.8	30.8	26.9	24.2	16.9	17.6	5.0	11.0	−10.5

$k = 44$									
s	*l*	*s*	*l*	*s*	*l*	*s*	*l*	*s*	*l*
10.8	—11.0	8.8	—17.1	6.8	—24.4	4.8	—33.6	2.8	—46.8
10.6	—11.6	8.6	—17.8	6.6	—25.2	4.6	—34.7	2.6	—48.6
10.4	—12.2	8.4	—18.5	6.4	—26.1	4.4	—35.8	2.4	—50.4
10.2	—12.8	8.2	—19.2	6.2	—26.9	4.2	—37.0	2.2	—52.4
10.0	—13.4	8.0	—19.9	6.0	—27.8	4.0	—38.2	2.0	—54.6
9.8	—14.0	7.8	—20.6	5.8	—28.7	3.8	—39.5	1.8	—56.9
9.6	—14.6	7.6	—21.3	5.6	—29.6	3.6	—40.8	1.6	—59.5
9.4	—15.2	7.4	—22.1	5.4	—30.6	3.4	—42.2	1.4	—62.4
9.2	—15.8	7.2	—22.8	5.2	—31.6	3.2	—43.7	1.2	—65.8
9.0	—16.5	7.0	—23.6	5.0	—32.6	3.0	—45.2	1.0	—69.6

k = 45									
s	*l*	*s*	*l*	*s*	*l*	*s*	*l*	*s*	*l*
45.0	45.0								
44.8	44.8	38.2	36.6	31.6	27.8	25.0	17.8	18.4	6.1
44.6	44.5	38.0	36.4	31.4	27.5	24.8	17.5	18.2	5.7
44.4	44.3	37.8	36.1	31.2	27.2	24.6	17.2	18.0	5.3
44.2	44.0	37.6	35.8			24.4	16.8	17.8	4.9
44.0	43.8	37.4	35.6	31.0	26.9	24.2	16.5	17.6	4.5
43.8	43.6	37.2	35.3	30.8	26.6	24.0	16.2	17.4	4.1
43.6	43.3			30.6	26.3	23.8	15.9	17.2	3.7
43.4	43.1	37.0	35.1	30.4	26.0	23.6	15.5		
43.2	42.8	36.8	34.8	30.2	25.8	23.4	15.2	17.0	3.3
		36.6	34.5	30.0	25.5	23.2	14.8	16.8	2.9
43.0	42.6	36.4	34.3	29.8	25.2			16.6	2.5
42.8	42.4	36.2	34.0	29.6	24.9	23.0	14.5	16.4	2.0
42.6	42.1	36.0	33.8	29.4	24.6	22.8	14.2	16.2	1.6
42.4	41.9	35.8	33.5	29.2	24.3	22.6	13.8	16.0	1.2
42.2	41.6	35.6	33.2			22.4	13.5	15.8	0.7
42.0	41.4	35.4	33.0	29.0	24.0	22.2	13.1	15.6	0.3
41.8	41.1	35.2	32.7	28.8	23.7	22.0	12.8	15.4	—0.1
41.6	40.9			28.6	23.4	21.8	12.4	15.2	—0.6
41.4	40.6	35.0	32.4	28.4	23.1	21.6	12.1		
41.2	40.4	34.8	32.2	28.2	22.8	21.4	11.7	15.0	—1.0
		34.6	31.9	28.0	22.5	21.2	11.4	14.8	—1.5
41.0	40.1	34.4	31.6	27.8	22.2			14.6	—1.9
40.8	39.9	34.2	31.4	27.6	21.9	21.0	11.0	14.4	—2.4
40.6	39.6	34.0	31.1	27.4	21.6	20.8	10.6	14.2	—2.9
40.4	39.4	33.8	30.8	27.2	21.3	20.6	10.3	14.0	—3.4
40.2	39.1	33.6	30.5			20.4	9.9	13.8	—3.8
40.0	38.9	33.4	30.3	27.0	21.0	20.2	9.5	13.6	—4.3
39.8	38.6	33.2	30.0	26.8	20.7	20.0	9.2	13.4	—4.8
39.6	38.4			26.6	20.4	19.8	8.8	13.2	—5.3
39.4	38.1	33.0	29.7	26.4	20.0	19.6	8.4		
39.2	37.9	32.8	29.4	26.2	19.7	19.4	8.0	13.0	—5.8
		32.6	29.2	26.0	19.4	19.2	7.7	12.8	—6.3
39.0	37.6	32.4	28.9	25.8	19.1			12.6	—6.8
38.8	37.4	32.2	28.6	25.6	18.8	19.0	7.3	12.4	—7.3
38.6	37.1	32.0	28.3	25.4	18.5	18.8	6.9	12.2	—7.8
38.4	36.9	31.8	28.0	25.2	18.1	18.6	6.5	12.0	—8.4

$k = 45$									
s	*l*	*s*	*l*	*s*	*l*	*s*	*l*	*s*	*l*
11.8	— 8.9	9.6	—15.3	7.4	—22.9	5.2	—32.5	3.0	—46.4
11.6	— 9.4	9.4	—15.9	7.2	—23.6			2.8	—48.0
11.4	—10.0	9.2	—16.5			5.0	—33.5	2.6	—49.8
11.2	—10.5			7.0	—24.4	4.8	—34.6	2.4	—51.7
		9.0	—17.2	6.8	—25.2	4.6	—35.7	2.2	—53.7
11.0	—11.1	8.8	—17.9	6.6	—26.1	4.4	—36.9	2.0	—55.9
10.8	—11.7	8.6	—18.5	6.4	—26.9	4 2	—38.0	1.8	—58.3
10.6	—12.3	8.4	—19.2	6.2	—27.8	4.0	—39.3	1.6	—61.0
10.4	—12.8	8.2	—19.9	6.0	—28.7	3.8	—40.6	1.4	—63.9
10.2	—13.4	8.0	—20.6	5.8	—29.6	3.6	—41.9	1.2	—67.3
10.0	—14.0	7.8	—21.4	5.6	—30.5	3.4	—43.3	1.0	—71.3
9.8	—14.7	7.6	—22.1	5.4	—31.5	3.2	—44.8		

$k = 46$

s	l	s	l	s	l	s	l	s	l
46.0	46.0								
45.8	45.8	39.2	37.7	32.6	28.9	26.0	19.1	19.4	7.6
45.6	45.5	39.0	37.4	32.4	28.6	25.8	18.8	19.2	7.2
45.4	45.3	38.8	37.2	32.2	28.3	25.6	18.4	19.0	6.8
45.2	45.0	38.6	36.9			25.4	18.1	18.8	6.4
45.0	44.8	38.4	36.6	32.0	28.0	25.2	17.8	18.6	6.1
44.8	44.6	38.2	36.4	31.8	27.8	25.0	17.5	18.4	5.7
44.6	44.3			31.6	27.5	24.8	17,1	18.2	5.3
44.4	44.1	38.0	36.1	31.4	27.2	24.6	16.8		
44.2	43.8	37.8	35.9	31.2	26.9	24.4	16.5	18.0	4.9
		37.6	35.6	31.0	26.6	24.2	16.1	17.8	4.5
44.0	43.6	37.4	35.3	30.8	26.3			17.6	4.0
43.8	43.4	37.2	35.1	30.6	26.0	24.0	15.8	17.4	3.6
43.6	43.1	37.0	34.8	30.4	25.7	25.8	15.5	17.2	3.2
43.4	42.9	36.8	34.6	30.2	25.5	23.6	15.1	17.0	2.8
43.2	42.7	36.6	34.3			23.4	14.8	16.8	2.4
43.0	42.4	36.4	34.0	30.0	25.2	23.2	14.5	16.6	2.0
42.8	42.2	36.2	33.8	29.8	24.9	23.0	14.1	16.4	1.5
42.6	41.9			29.6	24.6	22.8	13.8	16.2	1.1
42.4	41.7	36.0	33.5	29.4	24.3	22.6	13.4		
42.2	41.4	35.8	33.2	29.2	24.0	22.4	13.1	16.0	0.7
		35.6	33.0	29.0	23.7	22.2	12.7	15.8	0.2
42.0	41.2	35.4	32.7	28.8	23.4			15.6	—0.2
41.8	40.9	35.2	32.4	28.6	23.1	22.0	12.4	15.4	—0.7
41.6	40.7	35.0	32.2	28.4	22.8	21.8	12.0	15.2	—1.1
41.4	40.4	34.8	31.9	28.2	22.5	21.6	11.7	15.0	—1.6
41.2	40.2	34.6	31.6			21.4	11.3	14.8	—2.0
41.0	39.9	34.4	31.4	28.0	22.2	21.2	11.0	14.6	—2.5
40.8	39.7	34.2	31.1	27.8	21.9	21.0	10.6	14.4	—3.0
40.6	39.4			27.6	21.6	20.8	10.2	14.2	—3.4
40.4	39.2	34.0	30.8	27.4	21.3	20.6	9.9		
40.2	38.9	33.8	30.5	27.2	21.0	20.4	9.5	14.0	—3.9
		33.6	30.3	27.0	20.6	20.2	9.1	13.8	—4.4
40.0	38.7	33.4	30.0	26.8	20.3			13.6	—4.9
39.8	38.4	33.2	29.7	26.6	20.0	20.0	8.7	13.4	—5.4
39.6	38.2	33.0	29.4	26.4	19.7	19.8	8.4	13.2	—5.9
39.4	37.9	32.8	29.2	26.2	19.4	19.6	8.0	13.0	—6.4

$k = 46$									
s	l	s	l	s	l	s	l	s	l
12.8	— 6.9	10.4	—13.5	8.0	—21.4	5.6	—31.5	3.2	—46.0
12.6	— 7.4	10.2	—14.1	7.8	—22.2	5.4	—32.4	3.0	—47.6
12.4	— 7.9			7.6	—22.9	5.2	—33.5	2.8	—49.2
12.2	— 8.5	10.0	—14.7	7.4	—23.7	5.0	—34.5	2.6	—51.0
		9.8	—15.3	7.2	—24.5	4.8	—35.6	2.4	—53.0
12.0	— 9.0	9.6	—16.0	7.0	—25.3	4.6	—36.7	2.2	—55.0
11.8	— 9.5	9.4	—16.6	6.8	—26.1	4.4	—37.9		
11.6	—10.1	9.2	—17.3	6.6	—26.9	4.2	—39.1	2.0	—57.3
11.4	—10.6	9.0	—17.9	6.4	—27.8			1.8	—59.7
11.2	—11.2	8.8	—18.6	6.2	—28.7	4.0	—40.3	1.6	—62.4
11.0	—11.8	8.6	—19.3			3.8	—41.6	1.4	—65.4
10.8	—12.3	8.4	—20.0	6.0	—29.6	3.6	—43.0	1.2	—68.9
10.6	—12.9	8.2	—20.7	5.8	—30.5	3.4	—44.4	1.0	—72.9

k = 47

s	l	s	l	s	l	s	l	s	l
47.0	47.0								
46.8	46.8	40.2	38.7	33.6	30.0	27.0	20.3	20.4	9.1
46.6	46.5	40.0	38.5	33.4	29.7	26.8	20.0	20.2	8.7
46.4	46.3	39.8	38.2	33.2	29.4	26.6	19.7	20.0	8.3
46.2	46.1	39.6	37.9			26.4	19.4	19.8	7.9
46.0	45.8	39.4	37.7	33.0	29.2	26.2	19.0	19.6	7.5
45.8	45.6	39.2	37.4	32.8	28.9	26.0	18.7	19.4	7.2
45.6	45.3			32.6	28.6	25.8	18.4	19.2	6.8
45.4	45.1	39.0	37.2	32.4	28.3	25.6	18.1		
45.2	44.9	38.8	36.9	32.2	28.0	25.4	17.8	19.0	6.4
		38.6	36.7	32.0	27.8	25.2	17.4	18.8	6.0
45.0	44.6	38.4	36.4	31.8	27.5			18.6	5.6
44.8	44.4	38.2	36.2	31.6	27.2	25.0	17.1	18.4	5.2
44.6	44.1	38.0	35.9	31.4	26.9	24.8	16.8	18.2	4.8
44.4	43.9	37.8	35.6	31.2	26.6	24.6	16.4	18.0	4.4
44.2	43.7	37.6	35.4			24.4	16.1	17.8	4.0
44.0	43.4	37.4	35.1	31.0	26.3	24.2	15.8	17.6	3.6
43.8	43.2	37.2	34.8	30.8	26.0	24.0	15.4	17.4	3.2
43.6	42.9			30.6	25.7	23.8	15.1	17.2	2.7
43.4	42.7	37.0	34.6	30.4	25.5	23.6	14.8		
43.2	42.4	36.8	34.3	30.2	25.2	23.4	14.4	17.0	2.3
		36.6	34.1	30.0	24.9	23.2	14.1	16.8	1.9
43.0	42.2	36.4	33.8	29.8	24.6			16.6	1.5
42.8	41.9	36.2	33.5	29.6	24.3	23.0	13.7	16.4	1.0
42.6	41.7	36.0	33.3	29.4	24.0	22.8	13.4	16.2	0.6
42.4	41.5	35.8	33.0	29.2	23.7	22.6	13.0	16.0	0.2
42.2	41.2	35.6	32.7			22.4	12.7	15.8	−0.3
42.0	41.0	35.4	32.5	29.0	23.4	22.2	12.3	15.6	−0.7
41.8	40.7	35.2	32.2	28.8	23.1	22.0	12.0	15.4	−1.2
41.6	40.5			28.6	22.8	21.8	11.6	15.2	−1.6
41.4	40.2	35.0	31.9	28.4	22.5	21.6	11.3		
41.2	40.0	34.8	31.6	28.2	22.2	21.4	10.9	15.0	−2.1
		34.6	31.4	28.0	21.9	21.2	10.5	14.8	−2.6
41.0	39.7	34.4	31.1	27.8	21.6			14.6	−3.0
40.8	39.5	34.2	30.8	27.6	21.2	21.0	10.2	14.4	−3.5
40.6	39.2	34.0	30.6	27.4	20.9	20.8	9.8	14.2	−4.0
40.4	39.0	33.8	30.3	27.2	20.6	20.6	9.4	14.0	−4.5

$k = 47$									
s	*l*	*s*	*l*	*s*	*l*	*s*	*l*	*s*	*l*
13.8	— 5.0	11.4	—11.3	9.0	—18.7	6.2	—29.5	3.6	—44.1
13.6	— 5.5	11.2	—11.8	8.8	—19.3	6.0	—30.5	3.4	—45.6
13.4	— 6.0			8.6	—20.0	5.8	—31.4	3.2	—47.1
13.2	— 6.5	11.0	—12.4	8.4	—20.7	5.6	—32.4		
		10.8	—13.0	8.2	—21.5	5.4	—33.4	3.0	—48.7
13.0	— 7.0	10.6	—13.6	8.0	—22.2	5.2	—34.4	2.8	—50.4
12.8	— 7.5	10.4	—14.2	7.8	—22.9			2.6	—52.3
12.6	— 8.0	10.2	—14.8	7.6	—23.7	5.0	—35.5	2.4	—54.2
12.4	— 8.5	10.0	—15.4	7.4	—24.5	4.8	—36.6	2.2	—56.3
12.2	— 9.1	9.8	—16.0	7.2	—25.3	4.6	—37.7	2.0	—58.6
12.0	— 9.6	9.6	—16.7			4.4	—38.9	1.8	—61.1
11.8	—10.1	9.4	—17.3	7.0	—26.1	4.2	—40.1	1.6	—63.9
11.6	—10.7	9.2	—18.0	6.8	—26.9	4.0	—41.4	1.4	—66.9
				6.6	—27.8	3.8	—42.7	1.2	—70.5
				6.4	—28.6			1.0	—74.6

$k = 48$

s	*l*	*s*	*l*	*s*	*l*	*s*	*l*	*s*	*l*
48.0	48.0								
47.8	47.8	41.2	39.8	34.6	31.1	28.0	21.5	21.4	10.5
47.6	47.5	41.0	39.5	34.4	30.8	27.8	21.2	21.2	10.1
47.4	47.3	40.8	39.3	34.2	30.6	27.6	20.9	21.0	9.8
47.2	47.1	40.6	39.0			27.4	20.6	20.8	9.4
47.0	46.8	40.4	38.7	34.0	30.3	27.2	20.3	20.6	9.0
46.8	46.6	40.2	38.5	33.8	30.0	27.0	20.0	20.4	8.6
46.6	46.3			33.6	29.7	26.8	19.7	20.2	8.3
46.4	46.1	40.0	38.2	33.4	29.5	26.6	19.3		
46.2	45.9	39.8	38.0	33.2	29.2	26.4	19.0	20.0	7.9
		39.6	37.7	33.0	28.9	26.2	18.7	19.8	7.5
46.0	45.6	39.4	37.5	32.8	28.6			19.6	7.1
45.8	45.4	39.2	37.2	32.6	28.3	26.0	18.4	19.4	6.7
45.6	45.2	39.0	37.0	32.4	28.0	25.8	18.1	19.2	6.3
45.4	44.9	38.8	36.7	32.2	27.8	25.6	17.7	19.0	5.9
45.2	44.7	38.6	36.4			25.4	17.4	18.8	5.5
45.0	44.4	38.4	36.2	32.0	27.5	25.2	17.1	18.6	5.1
44.8	44.2	38.2	35.9	31.8	27.2	25.0	16.7	18.4	4.7
44.6	43.9			31.6	26.9	24.8	16.4	18.2	4.3
44.4	43.7	38.0	35.7	31.4	26.6	24.6	16.1		
44.2	43.5	37.8	35.4	31.2	26.3	24.4	15.7	18.0	3.9
		37.6	35.1	31.0	26.0	24.2	15.4	17.8	3.5
44.0	43.2	37.4	34.9	30.8	25.7			17.6	3.1
43.8	43.0	37.2	34.6	30.6	25.4	24.0	15.1	17.4	2.7
43.6	42.7	37.0	34.3	30.4	25.2	23.8	14.7	17.2	2.2
43.4	42.5	36.8	34.1	30.2	24.9	23.6	14.4	17.0	1.8
43.2	42.2	36.6	33.8			23.4	14.0	16.8	1.4
43.0	42.0	36.4	33.5	30.0	24.6	23.2	13.7	16.6	1.0
42.8	41.7	36.2	33.3	29.8	24.3	23.0	13.4	16.4	0.5
42.6	41.5			29.6	24.0	22.8	13.0	16.2	0.1
42.4	41.3	36.0	33.0	29.4	23.7	22.6	12.6		
42.2	41.0	35.8	32.7	29.2	23.4	22.4	12.3	16.0	−0.4
		35.6	32.5	29.0	23.1	22.2	11.9	15.8	−0.8
42.0	40.8	35.4	32.2	28.8	22.8			15.6	−1.3
41.8	40.5	35.2	31.9	28.6	22.5	22.0	11.6	15.4	−1.7
41.6	40.3	35.0	31.7	28.4	22.2	21.8	11.2	15.2	−2.2
41.4	40.0	34.8	31.4	28.2	21.8	21.6	10.9	15.0	−2.6

$k = 48$									
s	*l*	*s*	*l*	*s*	*l*	*s*	*l*	*s*	*l*
14.8	—3.1	12.0	—10.2	9.2	—18.7	6.4	—29.5	3.8	—43.8
14.6	—3.6	11.8	—10.8	9.0	—19.4	6.2	—30.4	3.6	—45.2
14.4	—4.1	11.6	—11.3	8.8	—20.1			3.4	—46.7
14.2	—4.5	11.4	—11.9	8.6	—20.8	6.0	—31.4	3.2	—48.3
		11.2	—12.5	8.4	—21.5	5.8	—32.3	3.0	—49.9
14.0	—5.0	11.0	—13.1	8.2	—22.2	5.6	—33.3	2.8	—51.6
13.8	—5.5	10.8	—13.6			5.4	—34.3	2.6	—53.5
13.6	—6.0	10.6	—14.2	8.0	—23.0	5.2	—35.4	2.4	—55.5
13.4	—6.5	10.4	—14.9	7.8	—23.7	5.0	—36.4	2.2	—57.6
13.2	—7.0	10.2	—15.5	7.6	—24.5	4.8	—37.6		
13.0	—7.6			7.4	—25.3	4.6	—38.7	2.0	—60.0
12.8	—8.1	10.0	—16.1	7.2	—26.1	4.4	—39.9	1.8	—62.5
12.6	—8.6	9.8	—16.7	7.0	—26.9	4.2	—41.2	1.6	—65.3
12.4	—9.1	9.6	—17.4	6.8	—27.8			1.4	—68.4
12.2	—9.7	9.4	—18.0	6.6	—28.6	4.0	—42.4	1.2	—72.0
								1.0	—76.2

$k = 49$									
s	*l*	*s*	*l*	*s*	*l*	*s*	*l*	*s*	*l*
49.0	49.0								
48.8	48.8	42.2	40.8	35.6	32.2	29.0	22.8	22.4	11.9
48.6	48.5	42.0	40.5	35.4	31.9	28.8	22.4	22.2	11.5
48.4	48.3	41.8	40.3	35.2	31.7	28.6	22.1	22.0	11.2
48.2	48.1	41.6	40.0			28.4	21.8	21.8	10.8
48.0	47.8	41.4	39.8	35.0	31.4	28.2	21.5	21.6	10.5
47.8	47.6	41.2	39.5	34.8	31.1	28.0	21.2	21.4	10.1
47.6	47.4			34.6	30.9	27.8	20.9	21.2	9.7
47.4	47.1	41.0	39.3	34.4	30.6	27.6	20.6		
47.2	46.9	40.8	39.0	34.2	30.3	27.4	20.3	21.0	9.3
		40.6	38.8	34.0	30.0	27.2	20.0	20.8	9.0
47.0	46.6	40.4	38.5	33.8	29.7			20.6	8.6
46.8	46.4	40.2	38.3	33.6	29.5	27.0	19.6	20.4	8.2
46.6	46.2	40.0	38.0	33.4	29.2	26.8	19.3	20.2	7.8
46.4	45.9	39.8	37.8	33.2	28.9	26.6	19.0	20.0	7.4
46.2	45.7	39.6	37.5			26.4	18.7	19.8	7.1
46.0	45.4	39.4	37.2	33.0	28.6	26.2	18.4	19.6	6.7
45.8	45.2	39.2	37.0	32.8	28.3	26.0	18.0	19.4	6.3
45.6	45.0			32.6	28.1	25.8	17.7	19.2	5.9
45.4	44.7	39.0	36.7	32.4	27.8	25.6	17.4		
45.2	44.5	38.8	36.5	32.2	27.5	25.4	17.1	19.0	5.5
		38.6	36.2	32.0	27.2	25.2	16.7	18.8	5.1
45.0	44.2	38.4	35.9	31.8	26.9			18.6	4.7
44.8	44.0	38.2	35.7	31.6	26.6	25.0	16.4	18.4	4.3
44.6	43.8	38.0	35.4	31.4	26.3	24.8	16.1	18.2	3.9
44.4	43.5	37.8	35.2	31.2	26.0	24.6	15.7	18.0	3.4
44.2	43.3	37.6	34.9			24.4	15.4	17.8	3.0
44.0	43.0	37.4	34.6	31.0	25.7	24.2	15.0	17.6	2.6
43.8	42.8	37.2	34.4	30.8	25.4	24.0	14.7	17.4	2.2
43.6	42.5			30.6	25.2	23.8	14.4	17.2	1.8
43.4	42.3	37.0	34.1	30.4	24.9	23.6	14.0		
43.2	42.0	36.8	33.8	30.2	24.6	23.4	13.7	17.0	1.3
		36.6	33.6	30.0	24.3	23.2	13.3	16.8	0.9
43.0	41.8	36.4	33.3	29.8	24.0			16.6	0.5
42.8	41.5	36.2	33.0	29.6	23.7	23.0	13.0	16.4	0.0
42.6	41.3	36.0	32.8	29.4	23.4	22.8	12.6	16.2	—0.4
42.4	41.0	35.8	32.5	29.2	23.1	22.6	12.3	16.0	—0.9

$k = 49$									
s	*l*	*s*	*l*	*s*	*l*	*s*	*l*	*s*	*l*
15.8	—1.3	13.0	— 8.1	10.0	—16.8	7.0	—27.7	4.0	—43.5
15.6	—1.8	12.8	— 8.7	9.8	—17.4	6.8	—28.6	3.8	—44.9
15.4	—2.2	12.6	— 9.2	9.6	—18.1	6.6	—29.5	3.6	—46.3
15.2	—2.7	12.4	— 9.7	9.4	—18.7	6.4	—30.4	3.4	—47.8
		12.2	—10.3	9.2	—19.4	6.2	—31.3	3.2	—49.4
15.0	—3.2	12.0	—10.8			6.0	—32.2		
14.8	—3.6	11.8	—11.4	9.0	—20.1	5.8	—33.2	3.0	—51.1
14.6	—4.1	11.6	—12.0	8.8	—20.8	5.6	—34.2	2.8	—52.9
14.4	—4.6	11.4	—12.5	8.6	—21.5	5.4	—35.2	2.6	—54.7
14.2	—5.1	11.2	—13.1	8.4	—22.2	5.2	—36.3	2.4	—56.8
14.0	—5.6			8.2	—23.0			2.2	—58.9
13.8	—6.1	11.0	—13.7	8.0	—23.7	5.0	—37.4	2.0	—61.3
13.6	—6.6	10.8	—14.3	7.8	—24.5	4.8	—38.5	1.8	—63.9
13.4	—7.1	10.6	—14.9	7.6	—25.3	4.6	—39.7	1.6	—66.7
13.2	—7.6	10.4	—15.5	7.4	—26.1	4.4	—40.9	1.4	—69.9
		10.2	—16.2	7.2	—26.9	4.2	—42.2	1.2	—73.6
								1.0	—77.9

k = 50									
s	*l*	*s*	*l*	*s*	*l*	*s*	*l*	*s*	*l*
50.0	50.0								
49.8	49.8	43.2	41.8	36.6	33.3	30.0	24.0	23.4	13.3
49.6	49.5	43.0	41.6	36.4	33.1	29.8	23.7	23.2	12.9
49.4	49.3	42.8	41.3	36.2	32.8	29.6	23.4	23.0	12.6
49.2	49.1	42.6	41.1			29.4	23.1	22.8	12.2
49.0	48.8	42.4	40.8	36.0	32.5	29.2	22.7	22.6	11.9
48.8	48.6	42.2	40.6	35.8	32.2	29.0	22.4	22.4	11.5
48.6	48.4			35.6	32.0	28.8	22.1	22.2	11.1
48.4	48.1	42.0	40.3	35.4	31.7	28.6	21.8		
48.2	47.9	41.8	40.1	35.2	31.4	28.4	21.5	22.0	10.8
		41.6	39.8	35.0	31.1	28.2	21.2	21.8	10.4
48.0	47.7	41.4	39.6	34.8	30.9			21.6	10.0
47.8	47.4	41.2	39.3	34.6	30.6	28.0	20.9	21.4	9.7
47.6	47.2	41.0	39.1	34.4	30.3	27.8	20.6	21.2	9.3
47.4	46.9	40.8	38.8	34.2	30.0	27.6	20.3	21.0	8.9
47.2	46.7	40.6	38.6			27.4	19.9	20.8	8.5
47.0	46.5	40.4	38.3	34.0	29.8	27.2	19.6	20.6	8.2
46.8	46.2	40.2	38.1	33.8	29.5	27.0	19.3	20.4	7.8
46.6	46.0			33.6	29.2	26.8	19.0	20.2	7.4
46.4	45.7	40.0	37.8	33.4	28.9	26.6	18.7		
46.2	45.5	39.8	37.5	33.2	28.6	26.4	18.3	20.0	7.0
		39.6	37.3	33.0	28.3	26.2	18.0	19.8	6.6
46.0	45.3	39.4	37.0	32.8	28.1			19.6	6.2
45.8	45.0	39.2	36.8	32.6	27.8	26.0	17.7	19.4	5.8
45.6	44.8	39.0	36.5	32.4	27.5	25.8	17.4	19.2	5.4
45.4	44.5	38.8	36.2	32.2	27.2	25.6	17.0	19.0	5.0
45.2	44.3	38.6	36.0			25.4	16.7	18.8	4.6
45.0	44.0	38.4	35.7	32.0	26.9	25.2	16.4	18.6	4.2
44.8	43.8	38.2	35.5	31.8	26.6	25.0	16.0	18.4	3.8
44.6	43.5			31.6	26.3	24.8	15.7	18.2	3.4
44.4	43.3	38.0	35.2	31.4	26.0	24.6	15.4		
44.2	43.1	37.8	34.9	31.2	25.7	24.4	15.0	18.0	3.0
		37.6	34.7	31.0	25.4	24.2	14.7	17.8	2.6
44.0	42.8	37.4	34.4	30.8	25.2			17.6	2.1
43.8	42.6	37.2	34.1	30.6	24.9	24.0	14.3	17.4	1.7
43.6	42.3	37.0	33.9	30.4	24.6	23.8	14.0	17.2	1.3
43.4	42.1	36.8	33.6	30.2	24.3	23.6	13.6	17.0	0.8

$k = 50$									
s	*l*	*s*	*l*	*s*	*l*	*s*	*l*	*s*	*l*
16.8	0.4	14.0	— 6.2	10.6	—15.6	7.4	—26.9	4.0	—44.6
16.6	—0.0	13.8	— 6.7	10.4	—16.2	7.2	—27.7	3.8	—46.0
16.4	—0.5	13.6	— 7.2	10.2	—16.8	7.0	—28.6	3.6	—47.4
16.2	—0.9	13.4	— 7.7			6.8	—29.4	3.4	—48.9
		13.2	— 8.2	10.0	—17.5	6.6	—30.3	3.2	—50.6
16.0	—1.4	13.0	— 8.7	9.8	—18.1	6.4	—31.2	3.0	—52.3
15.8	—1.8	12.8	— 9.3	9.6	—18.8	6.2	—32.2	2.8	—54.1
15.6	—2.3	12.6	— 9.8	9.4	—19.5			2.6	—56.0
15.4	—2.8	12.4	—10.3	9.2	—20.1	6.0	—33.1	2.4	—58.0
15.2	—3.2	12.2	—10.9	9.0	—20.8	5.8	—34.1	2.2	—60.2
15.0	—3.7			8.8	—21.5	5.6	—35.1		
14.8	—4.2	12.0	—11.5	8.6	—22.3	5.4	—36.2	2.0	—62.6
14.6	—4.7	11.8	—12.0	8.4	—23.0	5.2	—37.2	1.8	—65.3
14.4	—5.2	11.6	—12.6	8.2	—23.7	5.0	—38.4	1.6	—68.2
14.2	—5.7	11.4	—13.2			4.8	—39.5	1.4	—71.4
		11.2	—13.8	8.0	—24.5	4.6	—40.7	1.2	—75.2
		11.0	—14.4	7.8	—25.3	4.4	—41.9	1.0	—79.5
		10.8	—15.0	7.6	—26.1	4.2	—43.2		

$k = 52$									
s	*l*	*s*	*l*	*s*	*l*	*s*	*l*	*s*	*l*
52.0	52.0								
51.8	51.8	45.2	43.9	38.6	35.5	32.0	26.3	25.4	16.0
51.6	51.5	45.0	43.7	38.4	35.3	31.8	26.0	25.2	15.7
51.4	51.3	44.8	43.4	38.2	35.0	31.6	25.8	25.0	15.3
51.2	51.1	44.6	43.2			31.4	25.5	24.8	15.0
51.0	50.8	44.4	42.9	38.0	34.7	31.2	25.2	24.6	14.6
50.8	50.6	44.2	42.7	37.8	34.5	31.0	24.9	24.4	14.3
50.6	50.4			37.6	34.2	30.8	24.6	24.2	13.9
50.4	50.1	44.0	42.4	37.4	33.9	30.6	24.3		
50.2	49.9	43.8	42.2	37.2	33.6	30.4	24.0	24.0	13.6
		43.6	41.9	37.0	33.4	30.2	23.7	23.8	13.2
50.0	49.7	43.4	41.7	36.8	33.1			23.6	12.9
49.8	49.4	43.2	41.4	36.6	32.8	30.0	23.4	23.4	12.5
49.6	49.2	43.0	41.2	36.4	32.6	29.8	23.0	23.2	12.2
49.4	49.0	42.8	40.9	36.2	32.3	29.6	22.7	23.0	11.8
49.2	48.7	42.6	40.7			29.4	22.4	22.8	11.4
49.0	48.5	42.4	40.4	36.0	32.0	29.2	22.1	22.6	11.1
48.8	48.3	42.2	40.2	35.8	31.7	29.0	21.8	22.4	10.7
48.6	48.0			35.6	31.5	28.8	21.5	22.2	10.3
48.4	47.8	42.0	39.9	35.4	31.2	28.6	21.2		
48.2	47.5	41.8	39.7	35.2	30.9	28.4	20.9	22.0	10.0
		41.6	39.4	35.0	30.6	28.2	20.6	21.8	9.6
48.0	47.3	41.4	39.2	34.8	30.4			21.6	9.2
47.8	47.1	41.2	38.9	34.6	30.2	28.0	20.2	21.4	8.9
47.6	46.8	41.0	38.6	34.4	29.8	27.8	19.9	21.2	8.5
47.4	46.6	40.8	38.4	34.2	29.5	27.6	19.6	21.0	8.1
47.2	46.3	40.6	38.1			27.4	19.3	20.8	7.7
47.0	46.1	40.4	37.9	34.0	29.2	27.2	19.0	20.6	7.3
46.8	45.9	40.2	37.6	33.8	28.9	27.0	18.6	20.4	6.9
46.6	45.6			33.6	28.7	26.8	18.3	20.2	6.5
46.4	45.4	40.0	37.4	33.4	28.4	26.6	18.0		
46.2	45.1	39.8	37.1	33.2	28.1	26.4	17.7	20.0	6.1
		39.6	36.8	33.0	27.8	26.2	17.3	19.8	5.7
46.0	44.9	39.4	36.6	32.8	27.5			19.6	5.3
45.8	44.6	39.2	36.3	32.6	27.2	26.0	17.0	19.4	4.9
45.6	44.4	39.0	36.0	32.4	26.9	25.8	16.7	19.2	4.5
45.4	44.2	38.8	35.8	32.2	26.6	25.6	16.3	19.0	4.1

$k = 52$

s	*l*	*s*	*l*	*s*	*l*	*s*	*l*	*s*	*l*
18.8	3.7	15.2	−4.3	11.6	−13.9	8.0	−26.1	4.4	−44.0
18.6	3.3	15.0	−4.8	11.4	−14.4	7.8	−26.9	4.2	−45.3
18.4	2.9	14.8	−5.3	11.2	−15.0	7.6	−27.7		
18.2	2.5	14.6	−5.8	11.0	−15.7	7.4	−28.5	4.0	−46.7
		14.4	−6.3	10.8	−16.3	7.2	−29.4	3.8	−48.1
18.0	2.0	14.2	−6.8	10.6	−16.9	7.0	−30.2	3.6	−49.6
17.8	1.6			10.4	−17.5	6.8	−31.1	3.4	−51.2
17.6	1.2	14.0	−7.3	10.2	−18.2	6.6	−32.0	3.2	−52.8
17.4	0.7	13.8	−7.8			6.4	−33.0	3.0	−54.6
17.2	0.3	13.6	−8.3	10.0	−18.8	6.2	−33.9	2.8	−56.5
17.0	−0.1	13.4	−8.8	9.8	−19.5			2.6	−58.4
16.8	−0.6	13.2	−9.4	9.6	−20.2	6.0	−34.9	2.4	−60.6
16.6	−1.0	13.0	−9.9	9.4	−20.9	5.8	−35.9	2.2	−62.9
16.4	−1.5	12.8	−10.4	9.2	−21.6	5.6	−37.0		
16.2	−1.9	12.6	−11.0	9.0	−22.3	5.4	−38.0	2.0	−65.3
		12.4	−11.6	8.8	−23.0	5.2	−39.1	1.8	−68.1
16.0	−2.4	12.2	−12.1	8.6	−23.8	5.0	−40.3	1.6	−71.1
15.8	−2.9			8.4	−24.5	4.8	−41.5	1.4	−74.4
15.6	−3.3	12.0	−12.7	8.2	−25.3	4.6	−42.7	1.2	−78.3
15.4	−3.8	11.8	−13.3					1.0	−82.8

$k = 54$									
s	*l*	*s*	*l*	*s*	*l*	*s*	*l*	*s*	*l*
54.0	54.0								
53.8	53.8	47.2	46.0	40.6	37.7	34.0	28.7	27.4	18.6
53.6	53.5	47.0	45.7	40.4	37.4	33.8	28.4	27.2	18.3
53.4	53.3	46.8	45.5	40.2	37.2	33.6	28.1	27.0	18.0
53.2	53.1	46.6	45.3			33.4	27.8	26.8	17.6
53.0	52.8	46.4	45.0	40.0	36.9	33.2	27.5	26.6	17.3
52.8	52.6	46.2	44.8	39.8	36.7	33.0	27.2	26.4	17.0
52.6	52.4			39.6	36.4	32.8	27.0	26.2	16.6
52.4	52.2	46.0	44.5	39.4	36.1	32.6	26.7		
52.2	51.9	45.8	44.3	39.2	35.9	32.4	26.4	26.0	16.3
		45.6	44.0	39.0	35.6	32.2	26.1	25.8	16.0
52.0	51.7	45.4	43.8	38.8	35.3			25.6	15.6
51.8	51.5	45.2	43.5	38.6	35.1	32.0	25.8	25.4	15.3
51.6	51.2	45.0	43.3	38.4	34.8	31.8	25.5	25.2	14.9
51.4	51.0	44.8	43.0	38.2	34.5	31.6	25.2	25.0	14.6
51.2	50.8	44.6	42.8			31.4	24.9	24.8	14.2
51.0	50.5	44.4	42.5	38.0	34.3	31.2	24.6	24.6	14.0
50.8	50.3	44.2	42.3	37.8	34.0	31.0	24.3	24.4	13.5
50.6	50.0			37.6	33.7	30.8	24.0	24.2	13.2
50.4	49.8	44.0	42.0	37.4	33.4	30.6	23.7		
50.2	49.6	43.8	41.8	37.2	33.2	30.4	23.4	24.0	12.8
		43.6	41.5	37.0	32.9	30.2	23.1	23.8	12.5
50.0	49.3	43.4	41.3	36.8	32.6			23.6	12.1
49.8	49.1	43.2	41.0	36.6	32.3	30.0	22.8	23.4	11.8
49.6	48.9	43.0	40.8	36.4	32.1	29.8	22.4	23.2	11.4
49.4	48.6	42.8	40.5	36.2	31.8	29.6	22.1	23.0	11.0
49.2	48.4	42.6	40.3			29.4	21.8	22.8	10.7
49.0	48.2	42.4	40.0	36.0	31.5	29.2	21.5	22.6	10.3
48.8	47.9	42.2	39.8	35.8	31.2	29.0	21.2	22.4	9.9
48.6	47.7			35.6	31.0	28.8	20.9	22.2	9.6
48.4	47.4	42.0	39.5	35.4	30.7	28.6	20.6		
48.2	47.2	41.8	39.3	35.2	30.4	28.4	20.2	22.0	9.2
		41.6	39.0	35.0	30.1	28.2	19.9	21.8	8.8
48.0	47.0	41.4	38.7	34.8	29.8			21.6	8.4
47.8	46.7	41.2	38.5	34.6	29.6	28.0	19.6	21.4	8.0
47.6	46.5	41.0	38.2	34.4	29.3	27.8	19.3	21.2	7.6
47.4	46.2	40.8	38.0	34.2	29.0	27.6	19.0	21.0	7.3

$k = 54$									
s	*l*	*s*	*l*	*s*	*l*	*s*	*l*	*s*	*l*
20.8	6.9	16.8	— 1.6	12.8	—11.6	8.8	—24.5	4.8	—43.4
20.6	6.5	16.6	— 2.0	12.6	—12.2	8.6	—25.2	4.6	—44.7
20.4	6.1	16.4	— 2.5	12.4	—12.8	8.4	—26.0	4.4	—46.0
20.2	5.7	16.2	— 3.0	12.2	—13.3	8.2	—26.8	4.2	—47.4
20.0	5.3	16.0	— 3.4	12.0	—13.9	8.0	—27.6	4.0	—48.8
19.8	4.9	15.8	— 3.9	11.8	—14.5	7.8	—28.4	3.8	—50.3
19.6	4.5	15.6	— 4.4	11.6	—15.1	7.6	—29.3	3.6	—51.8
19.4	4.1	15.4	— 4.9	11.4	—15.7	7.4	—30.1	3.4	—53.4
19.2	3.6	15.2	— 5.4	11.2	—16.3	7.2	—31.0	3.2	—55.1
19.0	3.2	15.0	— 5.8	11.0	—17.0	7.0	—31.9	3.0	—57.0
18.8	2.8	14.8	— 6.3	10.8	—17.6	6.8	—32.8	2.8	—58.9
18.6	2.4	14.6	— 6.8	10.6	—18.2	6.6	—33.7	2.6	—60.9
18.4	2.0	14.4	— 7.4	10.4	—18.9	6.4	—34.7	2.4	—63.1
18.2	1.5	14.2	— 7.9	10.2	—19.5	6.2	—35.7	2.2	—65.5
18.0	1.1	14.0	— 8.4	10.0	—20.2	6.0	—36.7	2.0	—68.0
17.8	0.7	13.8	— 8.9	9.8	—20.9	5.8	—37.7	1.8	—70.8
17.6	0.2	13.6	— 9.4	9.6	—21.6	5.6	—38.8	1.6	—73.9
17.4	—0.2	13.4	—10.0	9.4	—22.3	5.4	—39.9	1.4	—77.4
17.2	—0.7	13.2	—10.5	9.2	—23.0	5.2	—40.0	1.2	—81.4
17.0	—1.1	13.0	—11.1	9.0	—23.7	5.0	—42.2	1.0	—86.1

$k = 56$									
s	*l*	*l*	*s*	*s*	*l*	*s*	*l*	*s*	*l*
56.0	56.0								
55.8	55.8	49.2	48.1	42.6	39.9	36.0	31.0	29.4	21.2
55.6	55.5	49.0	47.8	42.4	39.6	35.8	30.7	29.2	20.9
55.4	55.3	48.8	47.6	42.2	39.4	35.6	30.5	29.0	20.6
55.2	55.1	48.6	47.3			35.4	30.2	28.8	20.2
55.0	54.9	48.4	47.1	42.0	39.1	35.2	29.9	28.6	19.9
54.8	54.6	48.2	46.8	41.8	38.8	35.0	29.6	28.4	19.6
54.6	54.4			41.6	38.6	34.8	29.3	28.2	19.3
54.4	54.2	48.0	46.6	41.4	38.3	34.6	29.0		
54.2	53.9	47.8	46.4	41.2	38.1	34.4	28.7	28.0	19.0
		47.6	46.1	41.0	37.8	34.2	28.5	27.8	18.6
54.0	53.7	47.4	45.9	40.8	37.5			27.6	18.3
53.8	53.5	47.2	45.6	40.6	37.3	34.0	28.2	27.4	18.0
53.6	53.2	47.0	45.4	40.4	37.0	33.8	27.9	27.2	17.6
53.4	53.0	46.8	45.1	40.2	36.7	33.6	27.6	27.0	17.3
53.2	52.8	46.6	44.9			33.4	27.3	26.8	17.0
53.0	52.5	46.4	44.6	40.0	36.5	33.2	27.0	26.6	16.6
52.8	52.3	46.2	44.4	39.8	36.2	33.0	26.7	26.4	16.3
52.6	52.1			39.6	35.9	32.8	26.4	26.2	16.0
52.4	51.8	46.0	44.2	39.4	35.7	32.6	26.1		
52.2	51.6	45.8	43.9	39.2	35.4	32.4	25.8	26.0	15.6
		45.6	43.7	39.0	35.1	32.2	25.5	25.8	15.3
52.0	51.4	45.4	43.4	38.8	34.9			25.6	14.9
51.8	51.1	45.2	43.2	38.6	34.6	32.0	25.2	25.4	14.6
51.6	50.9	45.0	42.9	38.4	34.3	31.8	24.9	25.2	14.2
51.4	50.7	44.8	42.7	38.2	34.1	31.6	24.6	25.0	13.9
51.2	50.4	44.6	42.4			31.4	24.3	24.8	13.5
51.0	50.2	44.4	42.2	38.0	33.8	31.2	24.0	24.6	13.2
50.8	50.0	44.2	41.9	37.8	33.5	31.0	23.7	24.4	12.8
50.6	49.7			37.6	33.2	30.8	23.4	24.2	12.5
50.4	49.5	44.0	41.6	37.4	33.0	30.6	23.1		
50.2	49.2	43.8	41.4	37.2	32.7	30.4	22.8	24.0	12.1
		43.6	41.1	37.0	32.4	30.2	22.5	23.8	11.7
50.0	49.0	43.4	40.9	36.8	32.1			23.6	11.4
49.8	48.8	43.2	40.6	36.6	31.9	30.0	22.1	23.4	11.0
49.6	48.5	43.0	40.4	36.4	31.6	29.8	21.8	23.2	10.6
49.4	48.3	42.8	40.1	36.2	31.3	29.6	21.5	23.0	10.3

$k = 56$									
s	*l*	*s*	*l*	*s*	*l*	*s*	*l*	*s*	*l*
22.8	9.9	18.4	1.0	14.0	— 9.5	9.6	—23.0	5.2	—42.9
22.6	9.5	18.2	0.6	13.8	—10.0	9.4	—23.7	5.0	—44.1
22.4	9.1			13.6	—10.6	9.2	—24.5	4.8	—45.4
22.2	8.8	18.0	0.2	13.4	—11.1	9.0	—25.2	4.6	—46.7
		17.8	—0.3	13.2	—11.7	8.8	—26.0	4.4	—48.0
22.0	8.4	17.6	—0.7	13.0	—12.3	8.6	—26.7	4.2	—49.4
21.8	8.0	17.4	—1.2	12.8	—12.8	8.4	—27.5		
21.6	7.6	17.2	—1.6	12.6	—13.4	8.2	—28.3	4.0	—50.9
21.4	7.2	17.0	—2.1	12.4	—14.0			3.8	—52.4
21.2	6.8	16.8	—2.6	12.2	—14.6	8.0	—29.2	3.6	—54.0
21.0	6.4	16.6	—3.0			7.8	—30.0	3.4	—55.7
20.8	6.0	16.4	—3.5	12.0	—15.2	7.6	—30.9	3.2	—57.4
20.6	5.6	16.2	—4.0	11.8	—15.8	7.4	—31.7	3.0	—59.3
20.4	5.2			11.6	—16.4	7.2	—32.6	2.8	—61.3
20.2	4.8	16.0	—4.4	11.4	—17.0	7.0	—33.6	2.6	—63.4
		15.8	—4.9	11.2	—17.6	6.8	—34.5	2.4	—65.6
20.0	4.4	15.6	—5.4	11.0	—18.3	6.6	—35.4	2.2	—68.1
19.8	4.0	15.4	—5.9	10.8	—18.9	6.4	—36.4		
19.6	3.6	15.2	—6.4	10.6	—19.6	6.2	—37.4	2.0	—70.7
19.4	3.2	15.0	—6.9	10.4	—20.2			1.8	—73.6
19.2	2.8	14.8	—7.4	10.2	—20.9	6.0	—38.5	1.6	—76.8
19.0	2.3	14.6	—7.9			5.8	—39.5	1.4	—80.4
18.8	1.9	14.4	—8.5	10.0	—21.6	5.6	—40.6	1.2	—84.5
18.6	1.5	14.2	—9.0	9.8	—22.3	5.4	—41.8	1.0	—89.4

k = 58									
s	*l*	*s*	*l*	*s*	*l*	*s*	*l*	*s*	*l*
58.0	58.0								
57.8	57.8	51.2	50.1	44.6	42.0	38.0	33.3	31.4	23.7
57.6	57.5	51.0	49.9	44.4	41.8	37.8	33.0	31.2	23.4
57.4	57.3	50.8	49.6	44.2	41.5	37.6	32.8	31.0	23.1
57.2	57.1	50.6	49.4			37.4	32.5	30.8	22.8
57.0	56.9	50.4	49.2	44.0	41.3	37.2	32.2	30.6	22.5
56.8	56.6	50.2	48.9	43.8	41.0	37.0	31.9	30.4	22.2
56.6	56.4			43.6	40.7	36.8	31.6	30.2	21.9
56.4	56.2	50.0	48.7	43.4	40.5	36.6	31.4		
56.2	55.9	49.8	48.4	43.2	40.2	36.4	31.1	30.0	21.5
		49.6	48.2	43.0	40.0	36.2	30.8	29.8	21.2
56.0	55.7	49.4	48.0	42.8	39.7			29.6	20.9
55.8	55.5	49.2	47.7	42.6	39.5	36.0	30.5	29.4	20.6
55.6	55.3	49.0	47.5	42.4	39.2	35.8	30.2	29.2	20.3
55.4	55.0	48.8	47.2	42.2	38.9	35.6	29.9	29.0	19.9
55.2	54.8	48.6	47.0			35.4	29.7	28.8	19.6
55.0	54.6	48.4	46.7	42.0	38.7	35.2	29.4	28.6	19.3
54.8	54.3	48.2	46.5	41.8	38.4	35.0	29.1	28.4	19.0
54.6	54.1			41.6	38.2	34.8	28.8	28.2	18.6
54.4	53.9	48.0	46.3	41.4	37.9	34.6	28.5		
54.2	53.6	47.8	46.0	41.2	37.6	34.4	28.2	28.0	18.3
		47.6	45.8	41.0	37.4	34.2	27.9	27.8	18.0
54.0	53.4	47.4	45.5	40.8	37.1			27.6	17.6
53.8	53.2	47.2	45.3	40.6	36.8	34.0	27.6	27.4	17.3
53.6	52.9	47.0	45.0	40.4	36.6	33.8	27.3	27.2	17.0
53.4	52.7	46.8	44.8	40.2	36.3	33.6	27.0	27.0	16.6
53.2	52.5	46.6	44.5			33.4	26.7	26.8	16.3
53.0	52.2	46.4	44.3	40.0	36.0	33.2	26.5	26.6	16.0
52.8	52.0	46.2	44.0	39.8	35.8	33.0	26.2	26.4	15.6
52.6	51.8			39.6	35.5	32.8	25.9	26.2	15.3
52.4	51.5	46.0	43.8	39.4	35.2	32.6	25.6		
52.2	51.3	45.8	43.5	39.2	35.0	32.4	25.3	26.0	14.9
		45.6	43.3	39.0	34.7	32.2	24.9	25.8	14.6
52.0	51.1	45.4	43.0	38.8	34.4			25.6	14.2
51.8	50.8	45.2	42.8	38.6	34.1	32.0	24.6	25.4	13.9
51.6	50.6	45.0	42.5	38.4	33.9	31.8	24.3	25.2	13.5
51.4	50.4	44.8	42.3	38.2	33.6	31.6	24.0	25.0	13.2

$k = 58$

s	l	s	l	s	l	s	l	s	l
24.8	12.8	20.0	3.5	15.2	— 7.5	10.4	—21.6	5.8	—41.3
24.6	12.4	19.8	3.1	15.0	— 8.0	10.2	—22.3	5.6	—42.5
24.4	12.1	19.6	2.7	14.8	— 8.5			5.4	—43.6
24.2	11.7	19.4	2.3	14.6	— 9.0	10.0	—23.0	5.2	—44.8
		19.2	2.0	14.4	— 9.6	9.8	—23.7	5.0	—46.1
24.0	11.3	19.0	1.4	14.2	—10.1	9.6	—24.4	4.8	—47.4
23.8	11.0	18.8	1.0			9.4	—25.1	4.6	—48.7
23.6	10.6	18.6	0.6	14.0	—10.6	9.2	—25.9	4.4	—50.1
23.4	10.2	18.4	0.1	13.8	—11.2	9.0	—26.7	4.2	—51.5
23.2	9.7	18.2	— 0.3	13.6	—11.7	8.8	—27.4		
23.0	9.5			13.4	—12.3	8.6	—28.2	4.0	—53.0
22.8	9.1	18.0	—0.8	13.2	—12.9	8.4	—29.0	3.8	—54.6
22.6	8.7	17.8	—1.2	13.0	—13.4	8.2	—29.9	3.6	—56.2
22.4	8.3	17.6	—1.7	12.8	—14.0			3.4	—57.9
22.2	8.0	17.4	—2.1	12.6	—14.6	8.0	—30.7	3.2	—59.7
		17.2	—2.6	12.4	—15.2	7.8	—31.6	3.0	—61.7
22.0	7.6	17.0	—3.1	12.2	—15.8	7.6	—32.4	2.8	—63.7
21.8	7.2	16.8	—3.5			7.4	—33.3	2.6	—65.9
21.6	6.8	16.6	—4.0	12.0	—16.4	7.2	—34.3	2.4	—68.2
21.4	6.4	16.4	—4.5	11.8	—17.0	7.0	—35.2	2.2	—70.7
21.2	6.0	16.2	—5.0	11.6	—17.6	6.8	—36.1		
21.0	5.6			11.4	—18.3	6.6	—37.1	2.0	—73.4
20.8	5.2	16.0	—5.5	11.2	—18.9	6.4	—38.1	1.8	—76.4
20.6	4.8	15.8	—6.0	11.0	—19.6	6.2	—39.2	1.6	—79.7
10.4	4.4	15.6	—6.5	10.8	—20.2			1.4	—83.4
20.2	4.0	15.4	—7.0	10.6	—20.9	6.0	—40.2	1.2	—87.7
								1.0	—92.6

k = 60									
s	*l*	*s*	*l*	*s*	*l*	*s*	*l*	*s*	*l*
60.0	60.0								
59.8	59.8	53.2	52.2	46.6	44.2	40.0	35.6	33.4	26.2
59.6	59.5	53.0	51.9	46.4	43.9	39.8	35.3	33.2	25.9
59.4	59.3	52.8	51.7	46.2	43.7	39.6	35.0	33.0	25.6
59.2	59.1	52.6	51.5			39.4	34.8	32.8	25.3
59.0	58.9	52.4	51.2	46.0	43.4	39.2	34.5	32.6	25.0
58.8	58.6	52.2	51.0	45.8	43.2	39.0	34.2	32.4	24.7
58.6	58.4			45.6	42.9	38.8	34.0	32.2	24.4
58.4	58.2	52.0	50.8	45.4	42.7	38.6	33.7		
58.2	58.0	51.8	50.5	45.2	42.4	38.4	33.4	32.0	24.1
		51.6	50.3	45.0	42.1	38.2	33.1	31.8	23.8
58.0	57.7	51.4	50.0	44.8	41.9			31.6	23.5
57.8	57.5	51.2	49.8	44.6	41.6	38.0	32.8	31.4	23.2
57.6	57.3	51.0	49.6	44.4	41.4	37.8	32.6	31.2	22.8
57.4	57.1	50.8	49.3	44.2	41.1	37.6	32.3	31.0	22.5
57.2	56.8	50.6	49.1			37.4	32.0	30.8	22.2
57.0	56.6	50.4	48.8	44.0	40.9	37.2	31.7	30.6	21.9
56.8	56.4	50.2	48.6	43.8	40.6	37.0	31.4	30.4	21.6
56.6	56.1			43.6	40.3	36.8	31.2	30.2	21.3
56.4	55.9	50.0	48.3	43.4	40.1	36.6	30.9		
56.2	55.7	49.8	48.1	43.2	39.8	36.4	30.6	30.0	20.9
		49.6	47.9	43.0	39.6	36.2	30.3	29.8	20.6
56.0	55.4	49.4	47.6	42.8	39.3			29.6	20.3
55.8	55.2	49.2	47.4	42.6	39.0	36.0	30.0	29.4	20.0
55.6	55.0	49.0	47.1	42.4	38.8	35.8	29.7	29.2	19.6
55.4	54.7	48.8	46.9	42.2	38.5	35.6	29.4	29.0	19.3
55.2	54.5	48.6	46.6			35.4	29.2	28.8	19.0
55.0	54.3	48.4	46.4	42.0	38.3	35.2	28.9	28.6	18.7
54.8	54.0	48.2	46.2	41.8	38.0	35.0	28.6	28.4	18.3
54.6	53.8			41.6	37.7	34.8	28.3	28.2	18.0
54.4	53.6	48.0	45.9	41.4	37.5	34.6	28.0		
54.2	53.3	47.8	45.7	41.2	37.2	34.4	27.7	28.0	17.7
		47.6	45.4	41.0	36.9	34.2	27.4	27.8	17.3
54.0	53.1	47.4	45.2	40.8	36.7			27.6	17.0
53.8	52.9	47.2	44.9	40.6	36.4	34.0	27.1	27.4	16.6
53.6	52.6	47.0	44.7	40.4	36.1	33.8	26.8	27.2	16.3
53.4	52.4	46.8	44.4	40.2	35.9	33.6	26.5	27.0	16.0

$k = 60$									
s	*l*	*s*	*l*	*s*	*l*	*s*	*l*	*s*	*l*
26.8	15.6	21.6	6.0	16.2	—6.0	11.0	—20.9	6.0	—42.0
26.6	15.3	21.4	5.6			10.8	—21.5	5.8	—43.1
26.4	14.9	21.2	5.2	16.0	—6.5	10.6	—22.2	5.6	—44.3
26.2	14.6	21.0	4.8	15.8	—7.0	10.4	—22.9	5.4	—45.5
		20.8	4.4	15.6	—7.5	10.2	—23.6	5.2	—46.7
26.0	14.2	20.6	3.9	15.4	—8.0			5.0	—48.0
25.8	13.9	20.4	3.5	15.2	—8.5	10.0	—24.3	4.8	—49.3
25.6	13.5	20.2	3.1	15.0	—9.1	9.8	—25.1	4.6	—50.7
25.4	13.2			14.8	—9.6	9.6	—25.8	4.4	—52.1
25.2	12.8	20.0	2.7	14.6	—10.1	9.4	—26.6	4.2	—53.6
25.0	12.4	19.8	2.3	14.4	—10.7	9.2	—27.3		
24.8	12.1	19.6	1.8	14.2	—11.2	9.0	—28.1	4.0	—55.1
24.6	11.7	19.4	1.4			8.8	—28.9	3.8	—56.7
24.4	11.3	19.2	1.0	14.0	—11.7	8.6	—29.7	3.6	—58.4
24.2	11.0	19.0	0.5	13.8	—12.3	8.4	—30.5	3.4	—60.2
		18.8	0.1	13.6	—12.9	8.2	—31.4	3.2	—62.0
24.0	10.6	18.6	—0.4	13.4	—13.4			3.0	—64.0
23.8	10.2	18.4	—0.8	13.2	—14.0	8.0	—32.3	2.8	—66.1
23.6	9.9	18.2	—1.3	13.0	—14.6	7.8	—33.1	2.6	—68.3
23.4	9.5			12.8	—15.2	7.6	—34.0	2.4	—70.7
23.2	9.1	18.0	—1.7	12.6	—15.8	7.4	—34.9	2.2	—73.3
23.0	8.7	17.8	—2.2	12.4	—16.4	7.2	—35.9		
22.8	8.3	17.6	—2.6	12.2	—17.0	7.0	—36.8	2.0	—76.1
22.6	7.9	17.4	—3.1			6.8	—37.8	1.8	—79.2
22.4	7.6	17.2	—3.6	12.0	—17.6	6.6	—38.8	1.6	—82.6
22.2	7.2	17.0	—4.1	11.8	—18.3	6.4	—39.9	1.4	—86.4
		16.8	—4.5	11.6	—18.9	6.2	—40.9	1.2	—90.8
22.0	6.8	16.6	—5.0	11.4	—19.5			1.0	—95.9
21.8	6.4	16.4	—5.5	11.2	—20.2				

				$k = 62$					
s	*l*	*s*	*l*	*s*	*l*	*s*	*l*	*s*	*l*
62.0	62.0								
61.8	61.8	55.2	54.2	48.6	46.3	42.0	37.8	35.4	28.6
61.6	61.6	55.0	54.0	48.4	46.1	41.8	37.6	35.2	28.4
61.4	61.3	54.8	53.8	48.2	45.8	41.6	37.3	35.0	28.1
61.2	61.1	54.5	53.5			41.4	37.0	34.8	27.8
61.0	60.9	54.4	53.3	48.0	45.6	41.2	36.8	34.6	27.5
60.8	60.7	54.2	53.1	47.8	45.3	41.0	36.5	34.4	27.2
60.6	60.4			47.6	45.1	40.8	36.2	34.2	26.9
60.4	60.2	54.0	52.8	47.4	44.8	40.6	36.0		
60.2	60.0	53.8	52.6	47.2	44.6	40.4	35.7	34.0	26.6
		53.6	52.3	47.0	44.3	40.2	35.4	33.8	26.3
60.0	59.7	53.4	52.1	46.8	44.1			33.6	26.0
59.8	59.5	53.2	51.9	46.6	43.8	40.0	35.1	33.4	25.7
59.6	59.3	53.0	51.6	46.4	43.5	39.8	34.9	33.2	25.4
59.4	59.1	52.8	51.4	46.2	43.3	39.6	34.6	33.0	25.1
59.2	58.8	52.6	51.2			39.4	34.3	32.8	24.8
59.0	58.6	52.4	50.9	46.0	43.0	39.2	34.1	32.6	24.4
58.8	58.4	52.2	50.7	45.8	42.8	39.0	33.8	32.4	24.1
58.6	58.2			45.6	42.5	38.8	33.5	32.2	23.8
58.4	57.9	52.0	50.4	45.4	42.3	38.6	33.2		
58.2	57.7	51.8	50.2	45.2	42.0	38.4	32.9	32.0	23.5
		51.6	50.0	45.0	41.8	38.2	32.7	31.8	23.2
58.0	57.5	51.4	49.7	44.8	41.5			31.6	22.9
57.8	57.2	51.2	49.5	44.6	41.3	38.0	32.4	31.4	22.6
57.6	57.0	51.0	49.2	44.4	41.0	37.8	32.1	31.2	22.3
57.4	56.8	50.8	49.0	44.2	40.7	37.6	31.8	31.0	21.9
57.2	56.5	50.6	48.7			37.4	31.5	30.8	21.6
57.0	56.3	50.4	48.5	44.0	40.5	37.2	31.2	30.6	21.3
56.8	56.1	50.2	48.3	43.8	40.2	37.0	31.0	30.4	21.0
56.6	55.9			43.6	40.0	36.8	30.7	30.2	20.7
56.4	55.6	50.0	48.0	43.4	39.7	36.6	30.4		
56.2	55.4	49.8	47.8	43.2	39.4	36.4	30.1	30.0	20.3
		49.6	47.5	43.0	39.2	36.2	29.8	29.8	20.0
56.0	55.2	49.4	47.3	42.8	38.9			29.6	19.7
55.8	54.9	49.2	47.0	42.6	38.6	36.0	29.5	29.4	19.4
55.6	54.7	49.0	46.8	42.4	38.4	35.8	29.2	29.2	19.0
55.4	54.5	48.8	46.5	42.2	38.1	35.6	28.9	29.0	18.7

$k = 62$									
s	*l*	*s*	*l*	*s*	*l*	*s*	*l*	*s*	*l*
28.8	18.4	23.2	8.3	17.6	— 3.6	12.0	—18.9	6.4	—41.6
28.6	18.0	23.0	7.9	17.4	— 4.1	11.8	—19.5	6.2	—42.7
28.4	17.7	22.8	7.6	17.2	— 4.5	11.6	—20.2		
28.2	17.4	22.6	7.2	17.0	— 5.0	11.4	—20.8	6.0	—43.8
		22.4	6.8	16.8	— 5.5	11.2	—21.5	5.8	—45.0
28.0	17.0	22.2	6.4	16.6	— 6.0	11.0	—22.2	5.6	—46.1
27.8	16.7			16.4	— 6.5	10.8	—22.9	5.4	—47.4
27.6	16.3	22.0	6.0	16.2	— 7.0	10.6	—23.6	5.2	—48.6
27.4	16.0	21.8	5.6			10.4	—24.3	5.0	—49.9
27.2	15.6	21.6	5.2	16.0	— 7.5	10.2	—25.0	4.8	—51.3
27.0	15.3	21.4	4.8	15.8	— 8.0			4.6	—52.7
26.8	14.9	21.2	4.3	15.6	— 8.5	10.0	—25.7	4.4	—54.1
26.6	14.6	21.0	3.9	15.4	— 9.1	9.8	—26.5	4.2	—55.6
26.4	14.2	20.8	3.5	15.2	— 9.6	9.6	—27.2		
26.2	13.9	20.6	3.1	15.0	—10.1	9.4	—28.0	4.0	—57.2
		20.4	2.7	14.8	—10.7	9.2	—28.8	3.8	—58.9
26.0	13.5	20.2	2.2	14.6	—11.2	9.0	—29.6	3.6	—60.6
25.8	13.2			14.4	—11.8	8.8	—30.4	3.4	—62.4
25.6	12.8	20.0	1.8	14.2	—12.3	8.6	—31.2	3.2	—64.3
25.4	12.5	19.8	1.4			8.4	—32.1	3.0	—66.4
25.2	12.1	19.6	1.0	14.0	—12.9	8.2	—33.0	2.8	—68.5
25.0	11.7	19.4	0.5	13.8	—13.4			2.6	—70.8
24.8	11.4	19.2	0.1	13.6	—14.0	8.0	—33.8	2.4	—73.3
24.6	11.0	19.0	—0.4	13.4	—14.6	7.8	—34.7	2.2	—75.9
24.4	10.6	18.8	—0.8	13.2	—15.2	7.6	—35.6		
24.2	10.2	18.6	—1.2	13.0	—15.8	7.4	—36.6	2.0	—78.8
		18.4	—1.7	12.8	—16.4	7.2	—37.5	1.8	—82.0
24.0	9.9	18.2	—2.2	12.6	—17.0	7.0	—38.5	1.6	—85.5
23.8	9.5			12.4	—17.6	6.8	—39.5	1.4	—89.4
23.6	9.1	18.0	—2.7	12.2	—18.2	6.6	—40.5	1.2	—93.9
23.4	8.7	17.8	—3.1					1.0	—99.2

$k = 64$									
s	*l*	*s*	*l*	*s*	*l*	*s*	*l*	*s*	*l*
64.0	64.0								
63.8	63.8	57.2	56.3	50.6	48.4	44.0	40.1	37.4	31.1
63.6	63.6	57.0	56.0	50.4	48.2	43.8	39.8	37.2	30.8
63.4	63.3	56.8	55.8	50.2	47.9	43.6	39.6	37.0	30.5
63.2	63.1	56.6	55.6			43.4	39.3	36.8	30.2
63.0	62.9	56.4	55.3	50.0	47.7	43.2	39.0	36.6	29.9
62.8	62.7	56.2	55.1	49.8	47.4	43.0	38.8	36.4	29.6
62.6	62.4			49.6	47.2	42.8	38.5	36.2	29.3
62.4	62.2	56.0	54.9	49.4	46.9	42.6	38.2		
62.2	62.0	55.8	54.6	49.2	46.7	42.4	38.0	36.0	29.0
		55.6	54.4	49.0	46.5	42.2	37.7	35.8	28.7
62.0	61.8	55.4	54.2	48.8	46.2			35.6	28.4
61.8	61.5	55.2	53.9	48.6	46.0	42.0	37.4	35.4	28.1
61.6	61.3	55.0	53.7	48.4	45.7	41.8	37.2	35.2	27.8
61.4	61.1	54.8	53.5	48.2	45.5	41.6	36.9	35.0	27.5
61.2	60.9	54.6	53.2			41.4	36.6	34.8	27.2
61.0	60.6	54.4	53.0	48.0	45.2	41.2	36.3	34.6	26.9
60.8	60.4	54.2	52.8	47.8	45.0	41.0	36.1	34.4	26.6
60.6	60.2			47.6	44.7	40.8	35.8	34.2	26.3
60.4	60.0	54.0	52.5	47.4	44.5	40.6	35.5		
60.2	59.7	53.8	52.3	47.2	44.2	40.4	35.3	34.0	26.0
		53.6	52.0	47.0	43.9	40.2	35.0	33.8	25.7
60.0	59.5	53.4	51.8	46.8	43.7			33.6	25.4
59.8	59.3	53.2	51.6	46.6	43.4	40.0	34.7	33.4	25.1
59.6	59.0	53.0	51.3	46.4	43.2	39.8	34.4	33.2	24.8
59.4	58.8	52.8	51.1	46.2	42.9	39.6	34.2	33.0	24.5
59.2	58.6	52.6	50.8			39.4	33.9	32.8	24.2
59.0	58.4	52.4	50.6	46.0	42.7	39.2	33.6	32.6	23.9
58.8	58.1	52.2	50.4	45.8	42.4	39.0	33.3	32.4	23.6
58.6	57.9			45.6	42.2	38.8	33.0	32.2	23.3
58.4	57.7	52.0	50.1	45.4	41.9	38.6	32.8		
58.2	57.4	51.8	49.9	45.2	41.6	38.4	32.5	32.0	22.9
		51.6	49.6	45.0	41.4	38.2	32.2	31.8	22.6
58.0	57.2	51.4	49.4	44.8	41.1			31.6	22.3
57.8	57.0	51.2	49.2	44.6	40.9	38.0	31.9	31.4	22.0
57.6	56.7	51.0	48.9	44.4	40.6	37.8	31.6	31.2	21.7
57.4	56.5	50.8	48.7	44.2	40.3	37.6	31.3	31.0	21.4

$k = 64$

s	l	s	l	s	l	s	l	s	l
30.8	21.0	24.8	10.6	18.8	— 1.7	12.8	—17.6	6.8	— 41.2
30.6	20.7	24.6	10.3	18.6	— 2.2	12.6	—18.2	6.6	— 42.2
30.4	20.4	24.4	9.9	18.4	— 2.7	12.4	—18.8	6.4	— 43.3
30.2	20.1	24.2	9.5	18.2	— 3.1	12.2	—19.4	6.2	— 44.4
30.0	19.7	24.0	9.1	18.0	— 3.6	12.0	—20.0	6.0	— 45.6
29.8	19.4	23.8	8.7	17.8	— 4.1	11.8	—20.8	5.8	— 46.8
29.6	19.1	23.6	8.3	17.6	— 4.5	11.6	—21.4	5.6	— 48.0
29.4	18.7	23.4	8.0	17.4	— 5.0	11.4	—22.1	5.4	— 49.2
29.2	18.4	23.2	7.6	17.2	— 5.5	11.2	—22.8	5.2	— 50.5
29.0	18.1	23.0	7.2	17.0	— 6.0	11.0	—23.5	5.0	— 51.9
28.8	17.7	22.8	6.8	16.8	— 6.5	10.8	—24.2	4.8	— 53.2
28.0	17.4	22.0	0.4	10.0	— 7.0	10.0	—24.9	4.0	— 54.7
28.4	17.1	22.4	6.0	16.4	— 7.5	10.4	—25.6	4.4	— 56.2
28.2	16.7	22.2	5.6	16.2	— 8.0	10.2	—26.3	4.2	— 57.7
28.0	16.4	22.0	5.2	16.0	— 8.5	10.0	—27.0	4.0	— 59.3
27.8	16.0	21.8	4.8	15.8	— 9.1	9.8	—27.8	3.8	— 61.0
27.6	15.7	21.6	4.3	15.6	— 9.6	9.6	—28.6	3.6	— 62.8
27.4	15.3	21.4	3.9	15.4	—10.1	9.4	—29.4	3.4	— 64.7
27.2	15.0	21.2	3.5	15.2	—10.7	9.2	—30.2	3.2	— 66.6
27.0	14.6	21.0	3.1	15.0	—11.2	9.0	—31.0	3.0	— 68.7
26.8	14.3	20.8	2.7	14.8	—11.7	8.8	—31.9	2.8	— 70.9
26.6	13.9	20.6	2.2	14.6	—12.3	8.6	—32.7	2.6	— 73.3
26.4	13.6	20.4	1.8	14.4	—12.9	8.4	—33.6	2.4	— 75.8
26.2	13.2	20.2	1.4	14.2	—13.4	8.2	—34.4	2.2	— 78.5
26.0	12.8	20.0	1.0	14.0	—14.0	8.0	—35.3	2.0	— 81.5
25.8	12.5	19.8	0.5	13.8	—14.6	7.8	—36.3	1.8	— 84.8
25.6	12.1	19.6	0.1	13.6	—15.2	7.6	—37.2	1.6	— 88.4
25.4	11.7	19.4	—0.4	13.4	—15.7	7.4	—38.2	1.4	— 92.4
25.2	11.4	19.2	—0.8	13.2	—16.3	7.2	—39.1	1.2	— 97.1
25.0	11.0	19.0	—1.3	13.0	—16.9	7.0	—40.2	1.0	—102.5

$k = 66$									
s	*l*	*s*	*l*	*s*	*l*	*s*	*l*	*s*	*l*
66.0	66.0								
65.8	65.8	59.2	58.3	52.6	50.5	46.0	42.3	39.4	33.4
65.6	65.6	59.0	58.1	52.4	50.3	45.8	42.0	39.2	33.1
65.4	65.3	58.8	57.9	52.2	50.1	45.6	41.8	39.0	32.9
65.2	65.1	58.6	57.6			45.4	41.5	38.8	32.6
65.0	64.9	58.4	57.4	52.0	49.8	45.2	41.3	38.6	32.3
64.8	64.7	58.2	57.2	51.8	49.6	45.0	41.0	38.4	32.0
64.6	64.4			51.6	49.3	44.8	40.7	38.2	31.7
64.4	64.2	58.0	56.9	51.4	49.1	44.6	40.5		
64.2	64.0	57.8	56.7	51.2	48.8	44.4	40.2	38.0	31.4
		57.6	56.5	51.0	48.6	44.2	40.0	37.8	31.2
64.0	63.8	57.4	56.2	50.8	48.3			37.6	30.9
63.8	63.6	57.2	56.0	50.6	48.1	44.0	39.7	37.4	30.6
63.6	63.3	57.0	55.8	50.4	47.9	43.8	39.4	37.2	30.3
63.4	63.1	56.8	55.5	50.2	47.6	43.6	39.2	37.0	30.0
63.2	62.9	56.6	55.3			43.4	38.9	36.8	29.7
63.0	62.7	56.4	55.1	50.0	47.4	43.2	38.6	36.6	29.4
62.8	62.4	56.2	54.8	49.8	47.1	43.0	38.4	36.4	29.1
62.6	62.2			49.6	46.9	42.8	38.1	36.2	28.8
62.4	62.0	56.0	54.6	49.4	46.6	42.6	37.8		
62.2	61.7	55.8	54.4	49.2	46.4	42.4	37.6	36.0	28.5
		55.6	54.1	49.0	46.1	42.2	37.3	35.8	28.2
62.0	61.5	55.4	53.9	48.8	45.9			35.6	27.9
61.8	61.3	55.2	53.7	48.6	45.6	42.0	37.0	35.4	27.6
61.6	61.1	55.0	53.4	48.4	45.4	41.8	36.7	35.2	27.3
61.4	60.8	54.8	53.2	48.2	45.1	41.6	36.5	35.0	27.0
61.2	60.6	54.6	52.9			41.4	36.2	34.8	26.7
61.0	60.4	54.4	52.7	48.0	44.9	41.2	35.9	34.6	26.4
60.8	60.2	54.2	52.5	47.8	44.6	41.0	35.7	34.4	26.1
60.6	59.9			47.6	44.3	40.8	35.4	34.2	25.8
60.4	59.7	54.0	52.2	47.4	44.1	40.6	35.1		
60.2	59.5	53.8	52.0	47.2	43.8	40.4	34.8	34.0	25.5
		53.6	51.7	47.0	43.6	40.2	34.5	33.8	25.2
60.0	59.2	53.4	51.5	46.8	43.3			33.6	24.9
59.8	59.0	53.2	51.3	46.6	43.1	40.0	34.3	33.4	24.6
59.6	58.8	53.0	51.0	46.4	42.8	39.8	34.0	33.2	24.3
59.4	58.6	52.8	50.8	46.2	42.6	39.6	33.7	33.0	24.0

$k = 66$

s	*l*	*s*	*l*	*s*	*l*	*s*	*l*	*s*	*l*
32.8	23.6	26.4	12.9	20.0	0.1	13.6	−16.3	7.2	−40.8
32.6	23.3	26.2	12.5	19.8	−0.4	13.4	−16.9	7.0	−41.8
32.4	23.0			19.6	−0.8	13.2	−17.5	6.8	−42.9
32.2	22.7	26.0	12.1	19.4	−1.3	13.0	−18.1	6.6	−43.9
		25.8	11.8	19.2	−1.7	12.8	−18.7	6.4	−45.0
32.0	22.4	25.6	11.4	19.0	−2.2	12.6	−19.4	6.2	−46.2
31.8	22.1	25.4	11.0	18.8	−2.6	12.4	−20.0		
31.6	21.7	25.2	10.7	18.6	−3.1	12.2	−20.7	6.0	−47.4
31.4	21.4	25.0	10.3	18.4	−3.6			5.8	−48.6
31.2	21.1	24.8	9.9	18.2	−4.1	12.0	−21.3	5.6	−49.8
31.0	20.8	24.6	9.5			11.8	−22.0	5.4	−51.1
30.8	20.4	24.4	9.1	18.0	−4.5	11.6	−22.7	5.2	−52.4
30.6	20.1	24.2	8.8	17.8	−5.0	11.4	−23.4	5.0	−53.8
30.4	19.8			17.6	−5.5	11.2	−24.1	4.8	−55.2
30.2	19.5	24.0	8.4	17.4	−6.0	11.0	−24.8	4.6	−56.7
		23.8	8.0	17.2	−6.5	10.8	−25.5	4.4	−58.2
30.0	19.1	23.6	7.6	17.0	−7.0	10.6	−26.2	4.2	−59.8
29.8	18.8	23.4	7.2	16.8	−7.5	10.4	−26.9		
29.6	18.5	23.2	6.8	16.6	−8.0	10.2	−27.7	4.0	−61.4
29.4	18.1	23.0	6.4	16.4	−8.5			3.8	−63.2
29.2	17.8	22.8	6.0	16.2	−9.0	10.0	−28.5	3.6	−65.0
29.0	17.4	22.6	5.6			9.8	−29.2	3.4	−66.9
28.8	17.1	22.4	5.2	16.0	−9.6	9.6	−30.0	3.2	−68.9
28.6	16.8	22.2	4.8	15.8	−10.1	9.4	−30.8	3.0	−71.1
28.4	16.4			15.6	−10.6	9.2	−31.7	2.8	−73.3
28.2	16.1	22.0	4.4	15.4	−11.2	9.0	−32.5	2.6	−75.7
		21.8	4.0	15.2	−11.7	8.8	−33.3	2.4	−78.3
28.0	15.7	21.6	3.5	15.0	−12.2	8.6	−34.2	2.2	−81.1
27.8	15.4	21.4	3.1	14.8	−12.8	8.4	−35.1		
27.6	15.0	21.2	2.7	14.6	−13.4	8.2	−36.0	2.0	−84.2
27.4	14.7	21.0	2.3	14.4	−14.0			1.8	−87.5
27.2	14.3	20.8	1.8	14.2	−14.5	8.0	−36.9	1.6	−91.2
27.0	14.0	20.6	1.4			7.8	−37.8	1.4	−95.4
26.8	13.6	20.4	1.0	14.0	−15.1	7.6	−38.8	1.2	−100.2
26.6	13.2	20.2	0.5	13.8	−15.7	7.4	−39.8	1.0	−105.8

$k = 68$									
s	*l*	*s*	*l*	*s*	*l*	*s*	*l*	*s*	*l*
68.0	68.0								
67.8	67.8	61.2	60.4	54.6	52.7	48.0	44.5	41.4	35.8
67.6	67.6	61.0	60.1	54.4	52.4	47.8	44.3	41.2	35.5
97.4	67.3	60.8	59.9	54.2	52.1	47.6	44.0	41.0	35.2
67.2	67.1	60.6	59.7			47.4	43.7	40.8	34.9
67.0	66.9	60.4	59.5	54.0	51.9	47.2	43.5	40.6	34.7
66.8	66.7	60.2	59.2	53.8	51.7	47.0	43.2	40.4	34.4
66.6	66.5			53.6	51.4	46.8	43.0	40.2	34.1
66.4	66.2	60.0	59.0	53.4	51.2	46.6	42.7		
66.2	66.0	59.8	58.8	53.2	51.0	46.4	42.5	40.0	33.8
		59.6	58.5	53.0	50.7	46.2	42.2	39.8	33.5
66.0	65.8	59.4	58.3	52.8	50.5			39.6	33.3
65.8	65.6	59.2	58.1	52.6	50.2	46.0	41.9	39.4	33.0
65.6	65.3	59.0	57.8	52.4	50.0	45.8	41.7	39.2	32.7
65.4	65.1	58.8	57.6	52.2	49.7	45.6	41.4	39.0	32.4
65.2	64.9	58.6	57.4			45.4	41.1	38.8	32.1
65.0	64.7	58.4	57.1	52.0	49.5	45.2	40.9	38.6	31.8
64.8	64.4	58.2	56.9	51.8	49.3	45.0	40.6	38.4	31.6
64.6	64.2			51.6	49.0	44.8	40.3	38.2	31.3
64.4	64.0	58.0	56.7	51.4	48.8	44.6	40.1		
64.2	63.8	57.8	56.4	51.2	48.5	44.4	39.8	38.0	31.0
		57.6	56.2	51.0	48.3	44.2	39.6	37.8	30.7
64.0	63.5	57.4	56.0	50.8	48.0			37.6	30.4
63.8	63.3	57.2	55.7	50.6	47.8	44.0	39.3	37.4	30.1
63.6	63.1	57.0	55.5	50.4	47.5	43.8	39.0	37.2	29.8
63.4	62.9	56.8	55.3	50.2	47.3	43.6	38.8	37.0	29.5
63.2	62.6	56.6	55.0			43.4	38.5	36.8	29.2
63.0	62.4	56.4	54.8	50.0	47.0	43.2	38.2	36.6	28.9
62.8	62.2	56.2	54.6	49.8	46.8	43.0	38.0	36.4	28.6
62.6	62.0			49.6	46.5	42.8	37.7	36.2	28.3
62.4	61.7	56.0	54.3	49.4	46.3	42.6	37.4		
62.2	61.5	55.8	54.1	49.2	46.0	42.4	37.1	36.0	28.0
		55.6	53.8	49.0	45.8	42.2	36.9	35.8	27.7
62.0	61.3	55.4	53.6	48.8	45.5			35.6	27.4
61.8	61.1	55.2	53.4	48.6	45.3	42.0	36.6	35.4	27.1
61.6	60.8	55.0	53.1	48.4	45.0	41.8	36.3	35.2	26.8
61.4	60.6	54.8	52.9	48.2	44.8	41.6	36.1	35.0	26.5

$k = 68$

s	*l*	*s*	*l*	*s*	*l*	*s*	*l*	*s*	*l*
34.8	26.2	28.0	15.1	21.2	1.9	14.4	−15.0	7.8	−39.4
34.6	25.9	27.8	14.7	21.0	1.4	14.2	−15.6	7.6	−40.4
34.4	25.6	27.6	14.4	20.8	1.0			7.4	−41.4
34.2	25.3	27.4	14.0	20.6	0.6	14.0	−16.2	7.2	−42.4
		27.2	13.6	20.4	0.1	13.8	−16.8	7.0	−43.5
34.0	25.0	27.0	13.3	20.2	−0.3	13.6	−17.4	6.8	−44.5
33.8	24.7	26.8	12.9			13.4	−18.0	6.6	−45.6
33.6	24.4	26.6	12.6	20.0	−0.8	13.2	−18.7	6.4	−46.8
33.4	24.0	26.4	12.2	19.8	−1.2	13.0	−19.3	6.2	−47.9
33.2	23.7	26.2	11.8	19.6	−1.7	12.8	−19.9		
33.0	23.4			19.4	−2.1	12.6	−20.6	6.0	−49.1
32.8	23.1	26.0	11.5	19.2	−2.6	12.4	−21.2	5.8	−50.4
32.6	22.8	25.8	11.1	19.0	−3.1	12.2	−21.9	5.6	−51.6
32.4	22.5	25.6	10.7	18.8	−3.5			5.4	−53.0
32.2	22.1	25.4	10.3	18.6	−4.0	12.0	−22.6	5.2	−54.3
		25.2	9.9	18.4	−4.5	11.8	−23.2	5.0	−55.7
32.0	21.8	25.0	9.6	18.2	−5.0	11.6	−23.9	4.8	−57.2
31.8	21.5	24.8	9.2			11.4	−24.6	4.6	−58.7
31.6	21.2	24.6	8.8	18.0	−5.5	11.2	−25.3	4.4	−60.2
31.4	20.8	24.4	8.4	17.8	−6.0	11.0	−26.1	4.2	−61.9
31.2	20.5	24.2	8.0	17.6	−6.5	10.8	−26.8		
31.0	20.2			17.4	−7.0	10.6	−27.5	4.0	−63.6
30.8	19.9	24.0	7.6	17.2	−7.5	10.4	−28.3	3.8	−65.3
30.6	19.5	23.8	7.2	17.0	−8.0	10.2	−29.1	3.6	−67.2
30.4	19.2	23.6	6.8	16.8	−8.5			3.4	−69.2
30.2	18.9	23.4	6.4	16.6	−9.0	10.0	−29.8	3.2	−71.2
		23.2	6.0	16.4	−9.5	9.8	−30.6	3.0	−73.4
30.0	18.5	23.0	5.6	16.2	−10.1	9.6	−31.4	2.8	−75.7
29.8	18.2	22.8	5.2			9.4	−32.3	2.6	−78.2
29.6	17.8	22.6	4.8	16.0	−10.6	9.2	−33.1	2.4	−80.9
29.4	17.5	22.4	4.4	15.8	−11.1	9.0	−33.9	2.2	−83.8
29.2	17.2	22.2	4.0	15.6	−11.7	8.8	−34.8		
29.0	16.8			15.4	−12.2	8.6	−35.7	2.0	−86.9
28.8	16.5	22.0	3.6	15.2	−12.8	8.4	−36.6	1.8	−90.3
28.6	16.1	21.8	3.1	15.0	−13.3	8.2	−37.5	1.6	−94.1
28.4	15.8	21.6	2.7	14.8	−13.9			1.4	−98.4
28.2	15.4	21.4	2.3	14.6	−14.5	8.0	−38.4	1.2	−103.3
								1.0	−109.1

$k = 70$									
s	l	s	l	s	l	s	l	s	l
70.0	70.0								
69.8	69.8	63.2	62.4	56.6	54.8	50.0	46.7	43.4	38.1
69.6	69.6	63.0	62.2	56.4	54.5	49.8	46.4	43.2	37.8
69.4	69.3	62.8	62.0	56.2	54.3	49.6	46.2	43.0	37.6
69.2	69.1	62.6	61.7			49.4	45.9	42.8	37.3
69.0	68.9	62.4	61.5	56.0	54.0	49.2	45.7	42.6	37.0
68.8	68.7	62.2	61.3	55.8	53.8	49.0	45.4	42.4	36.7
68.6	68.5			55.6	53.6	48.8	45.2	42.2	36.5
68.4	68.2	62.0	61.0	55.4	53.3	48.6	44.9		
68.2	68.0	61.8	60.8	55.2	53.1	48.4	44.7	42.0	36.2
		61.6	60.6	55.0	52.8	48.2	44.4	41.8	35.9
68.0	67.8	61.4	60.4	54.8	52.6			41.6	35.6
67.8	67.6	61.2	60.1	54.6	52.4	48.0	44.2	41.4	35.4
67.6	67.4	61.0	59.9	54.4	52.1	47.8	43.9	41.2	35.1
67.4	67.1	60.8	59.7	54.2	51.9	47.6	43.6	41.0	34.8
67.2	66.9	60.6	59.4			47.4	43.4	40.8	34.5
67.0	66.7	60.4	59.2	54.0	51.6	47.2	43.1	40.6	34.2
66.8	66.5	60.2	59.0	53.8	51.4	47.0	42.9	40.4	34.0
66.6	66.2			53.6	51.2	46.8	42.6	40.2	33.7
66.4	66.0	60.0	58.7	53.4	50.9	46.6	42.3		
66.2	65.8	59.8	58.5	53.2	50.7	46.4	42.1	40.0	33.4
		59.6	58.3	53.0	50.4	46.2	41.8	39.8	33.1
66.0	65.6	59.4	58.0	52.8	50.2			39.6	32.8
65.8	65.3	59.2	57.8	52.6	49.9	46.0	41.6	39.4	32.5
65.6	65.1	59.0	57.6	52.4	49.7	45.8	41.3	39.2	32.2
65.4	64.9	58.8	57.3	52.2	49.4	45.6	41.0	39.0	32.0
65.2	64.7	58.6	57.1			45.4	40.8	38.8	31.7
65.0	64.5	58.4	56.9	52.0	49.2	45.2	40.5	38.6	31.4
64.8	64.2	58.2	56.6	51.8	48.9	45.0	40.2	38.4	31.1
64.6	64.0			51.6	48.7	44.8	40.0	38.2	30.8
64.4	63.8	58.0	56.4	51.4	48.4	44.6	39.7		
64.2	63.5	57.8	56.2	51.2	48.2	44.4	39.4	38.0	30.5
		57.6	55.9	51.0	48.0	44.2	39.2	37.8	30.2
64.0	63.3	57.4	55.7	50.8	47.7			37.6	29.9
63.8	63.1	57.2	55.5	50.6	47.5	44.0	38.9	37.4	29.6
63.6	62.9	57.0	55.2	50.4	47.2	43.8	38.6	37.2	29.3
63.4	62.6	56.8	55.0	50.2	47.0	43.6	38.4	37.0	29.0

$k = 70$

s	l	s	l	s	l	s	l	s	l
36.8	28.7	29.6	17.2	22.2	3.2	15.2	−13.8	8.0	−40.0
36.6	28.4	29.4	16.9			15.0	−14.4	7.8	−41.0
36.4	28.1	29.2	16.5	22.0	2.8	14.8	−15.0	7.6	−42.0
36.2	27.8	29.0	16.2	21.8	2.3	14.6	−15.6	7.4	−43.0
		28.8	15.8	21.6	1.9	14.4	−16.1	7.2	−44.0
36.0	27.5	28.6	15.5	21.4	1.5	14.2	−16.7	7.0	−45.1
35.8	27.2	28.4	15.1	21.2	1.0			6.8	−46.2
35.6	26.9	28.2	14.8	21.0	0.6	14.0	−17.3	6.6	−47.3
35.4	26.6			20.8	0.2	13.8	−18.0	6.4	−48.5
35.2	26.3	28.0	14.4	20.6	−0.3	13.6	−18.6	6.2	−49.7
35.0	26.0	27.8	14.1	20.4	−0.7	13.4	−19.2		
34.8	25.7	27.6	13.7	20.2	−1.2	13.2	−19.8	6.0	−50.9
34.6	25.4	27.4	13.3			13.0	−20.5	5.8	−52.2
34.4	25.1	27.2	13.0	20.0	−1.6	12.8	−21.1	5.6	−53.5
34.2	24.8	27.0	12.6	19.8	−2.1	12.6	−21.8	5.4	−54.8
		26.8	12.3	19.6	−2.6	12.4	−22.4	5.2	−56.2
34.0	24.5	26.6	11.9	19.4	−3.0	12.2	−23.1	5.0	−57.6
33.8	24.1	26.4	11.5	19.2	−3.5			4.8	−59.1
33.6	23.8	26.2	11.1	19.0	−4.0	12.0	−23.8	4.6	−60.7
33.4	23.5			18.8	−4.5	11.8	−24.5	4.4	−62.3
33.2	23.2	26.0	10.8	18.6	−4.9	11.6	−25.2	4.2	−63.9
33.0	22.9	25.8	10.4	18.4	−5.4	11.4	−25.9		
32.8	22.5	25.6	10.0	18.2	−5.9	11.2	−26.6	4.0	−65.7
32.6	22.2	25.4	9.6			11.0	−27.4	3.8	−67.5
32.4	21.9	25.2	9.2	18.0	−6.4	10.8	−28.1	3.6	−69.4
32.2	21.6	25.0	8.9	17.8	−6.9	10.6	−28.9	3.4	−71.4
		24.8	8.5	17.6	−7.4	10.4	−29.6	3.2	−73.5
32.0	21.3	24.6	8.1	17.4	−7.9	10.2	−30.4	3.0	−75.8
31.8	20.9	24.4	7.7	17.2	−8.4			2.8	−78.1
31.6	20.6	24.2	7.3	17.0	−8.9	10.0	−31.2	2.6	−80.7
31.4	20.3			16.8	−9.5	9.8	−32.0	2.4	−83.4
31.2	19.9	24.0	6.9	16.6	−10.0	9.6	−32.8	2.2	−86.4
31.0	19.6	23.8	6.5	16.4	−10.5	9.4	−33.7		
30.8	19.3	23.6	6.1	16.2	−11.1	9.2	−34.5	2.0	−89.6
30.6	18.9	23.4	5.7			9.0	−35.4	1.8	−93.1
30.4	18.6	23.2	5.3	16.0	−11.6	8.8	−36.3	1.6	−97.0
30.2	18.3	23.0	4.9	15.8	−12.2	8.6	−37.2	1.4	−101.4
		22.8	4.4	15.6	−12.7	8.4	−38.1	1.2	−106.4
30.0	17.9	22.6	4.0	15.4	−13.3	8.2	−39.0	1.0	−112.4
29.8	17.6	22.4	3.6						

$k = 75$									
s	*l*	*s*	*l*	*s*	*l*	*s*	*l*	*s*	*l*
75.0	75.0								
74.8	74.8	67.6	66.8	60.6	58.8	53.4	50.2	46.4	41.2
74.6	74.6	67.4	66.6	60.4	58.6	53.2	49.9	46.2	40.9
74.4	74.3	67.2	66.4	60.2	58.4			46.0	40.6
74.2	74.1			60.0	58.1	53.0	49.7	45.8	40.4
74.0	73.9	67.0	66.2	59.8	57.9	52.8	49.4	45.6	40.1
73.8	73.7	66.8	65.9	59.6	57.6	52.6	49.2	45.4	39.8
73.6	73.5	66.6	65.7	59.4	57.4	52.4	48.9	45.2	39.6
73.4	73.3	66.4	65.5	59.2	57.2	52.2	48.7		
73.2	73.0	66.2	65.3			52.0	48.4	45.0	39.3
		66.0	65.0	59.0	56.9	51.8	48.2	44.8	39.0
73.0	72.8	65.8	64.8	58.8	56.7	51.6	47.9	44.6	38.7
72.8	72.6	65.6	64.6	58.6	56.5	51.4	47.7	44.4	38.5
72.6	72.4	65.4	64.4	58.4	56.2	51.2	47.4	44.2	38.2
72.4	72.2	65.2	64.1	58.2	56.0			44.0	37.9
72.2	71.9			58.0	55.7	51.0	47.1	43.8	37.7
72.0	71.7	65.0	63.9	57.8	55.5	50.8	46.9	43.6	37.4
71.8	71.5	64.8	63.7	57.6	55.3	50.6	46.6	43.4	37.1
71.6	71.3	64.6	63.4	57.4	55.0	50.4	46.4	43.2	36.8
71.4	71.1	64.4	63.2	57.2	54.8	50.2	46.1		
71.2	70.8	64.2	63.0			50.0	45.9	43.0	36.5
		64.0	62.8	57.0	54.5	49.8	45.6	42.8	36.3
71.0	70.6	63.8	62.5	56.8	54.3	49.6	45.4	42.6	36.0
70.8	70.4	63.6	62.3	56.6	54.1	49.4	45.1	42.4	35.7
70.6	70.2	63.4	62.1	56.4	53.8	49.2	44.8	42.2	35.4
70.4	70.0	63.2	61.8	56.2	53.6			42.0	35.1
70.2	69.7			56.0	53.3	49.0	44.6	41.8	34.9
70.0	69.5	63.0	61.6	55.8	53.1	48.8	44.3	41.6	34.6
69.8	69.3	62.8	61.4	55.6	52.9	48.6	44.1	41.4	34.3
69.6	69.1	62.6	61.1	55.4	52.6	48.4	43.8	41.2	34.0
69.4	68.9	62.4	60.9	55.2	52.4	48.2	43.5		
69.2	68.6	62.2	60.7			48.0	43.3	41.0	33.7
		62.0	60.5	55.0	52.1	47.8	43.0	40.8	33.4
69.0	68.4	61.8	60.2	54.8	51.9	47.6	42.8	40.6	33.2
68.8	68.2	61.6	60.0	54.6	51.6	47.4	42.5	40.4	32.9
68.6	68.0	61.4	59.8	54.4	51.4	47.2	42.2	40.2	32.6
68.4	67.7	61.2	59.5	54.2	51.1			40.0	32.3
68.2	67.5			54.0	50.9	47.0	42.0	39.8	32.0
68.0	67.3	61.0	59.3	53.8	50.7	46.8	41.7	39.6	31.7
67.8	67.1	60.8	59.1	53.6	50.4	46.6	41.4	39.4	31.4
								39.2	31.1

$k = 75$

s	l	s	l	s	l	s	l	s	l
39.0	30.8	31.2	18.5	23.6	4.2	16.0	−14.2	8.4	− 41.9
38.8	30.5			23.4	3.8	15.8	−14.7	8.2	− 42.9
38.6	30.2	31.0	18.1	23.2	3.4	15.6	−15.3	8.0	− 43.9
38.4	30.0	30.8	17.8			15.4	−15.9	7.8	− 44.9
38.2	29.6	30.6	17.5	23.0	2.9	15.2	−16.5	7.6	− 45.9
38.0	29.3	30.4	17.1	22.8	2.5			7.4	− 47.0
37.8	29.0	30.2	16.8	22.6	2.1	15.0	−17.1	7.2	− 48.1
37.6	28.7	30.0	16.4	22.4	1.6	14.8	−17.7		
37.4	28.4	29.8	16.1	22.2	1.2	14.6	−18.3	7.0	− 49.3
37.2	28.1	29.6	15.7	22.0	0.8	14.4	−18.9	6.8	− 50.4
		29.4	15.3	21.8	0.3	14.2	−19.5	6.6	− 51.6
37.0	27.8	29.2	15.0	21.6	− 0.1	14.0	−20.1	6.4	− 52.8
36.8	27.5			21.4	− 0.6	13.8	−20.8	6.2	− 54.1
36.6	27.2	29.0	14.6	21.2	− 1.0	13.6	−21.4	6.0	− 55.4
36.4	26.9	28.8	14.3			13.4	−22.1	5.8	− 56.7
36.2	26.6	28.6	13.9	21.0	− 1.5	13.2	−22.7	5.6	− 58.1
36.0	26.3	28.4	13.5	20.8	− 1.9			5.4	− 59.5
35.8	26.0	28.2	13.2	20.6	− 2.4	13.0	−23.4	5.2	− 60.9
35.6	25.7	28.0	12.8	20.4	− 2.9	12.8	−24.1		
35.4	25.3	27.8	12.4	20.2	− 3.3	12.6	−24.8	5.0	− 62.5
35.2	25.0	27.6	12.1	20.0	− 3.8	12.4	−25.5	4.8	− 64.0
		27.4	11.7	19.8	− 4.3	12.2	−26.2	4.6	− 65.6
35.0	24.7	27.2	11.3	19.6	− 4.8	12.0	−26.9	4.4	− 67.3
34.8	24.4			19.4	− 5.3	11.8	−27.6	4.2	− 69.1
34.6	24.1	27.0	10.9	19.2	− 5.7	11.6	−28.3	4.0	− 70.9
34.4	23.8	26.8	10.6			11.4	−29.1	3.8	− 72.9
34.2	23.4	26.6	10.2	19.0	− 6.2	11.2	−29.8	3.6	− 74.9
34.0	23.1	26.4	9.8	18.8	− 6.7			3.4	− 77.0
33.8	22.8	26.2	9.4	18.6	− 7.2	11.0	−30.6	3.2	− 79.3
33.6	22.5	26.0	9.0	18.4	− 7.7	10.8	−31.4		
33.4	22.2	25.8	8.6	18.2	− 8.2	10.6	−32.2	3.0	− 81.6
33.2	21.8	25.6	8.2	18.0	− 8.8	10.4	−33.0	2.8	− 84.2
		25.4	7.9	17.8	− 9.3	10.2	−33.8	2.6	− 86.9
33.0	21.5	25.2	7.5	17.6	− 9.8	10.0	−34.6	2.4	− 89.8
32.8	21.2			17.4	−10.3	9.8	−35.5	2.2	− 92.9
32.6	20.8	25.0	7.1	17.2	−10.9	9.6	−36.4	2.0	− 96.3
32.4	20.5	24.8	6.7			9.4	−37.2	1.8	−100.1
32.2	20.2	24.6	6.3	17.0	−11.4	9.2	−38.1	1.6	−104.2
32.0	19.8	24.4	5.8	16.8	−11.9			1.4	−108.9
31.8	19.5	24.2	5.4	16.6	−12.5	9.0	−39.0	1.2	−114.3
31.6	19.2	24.0	5.0	16.4	−13.0	8.8	−40.0	1.0	−120.6
31.4	18.8	23.8	4.6	16.2	−13.6	8.6	−40.9		

$k = 80$

s	l	s	l	s	l	s	l	s	l
80.0	80.0								
79.8	79.8	72.0	71.3	64.0	62.2	56.4	53.1	48.8	43.5
79.6	79.6	71.8	71.0	63.8	62.0	56.2	52.9	48.6	43.2
79.4	79.4	71.6	70.8	63.6	61.7			48.4	42.9
79.2	79.1	71.4	70.6	63.4	61.5	56.0	52.6	48.2	42.7
79.0	78.9	71.2	70.4	63.2	61.3	55.8	52.4		
78.8	78.7	71.0	70.2	63.0	61.0	55.6	52.1	48.0	42.4
78.6	78.5	70.8	69.9	62.8	60.8	55.4	51.9	47.8	42.1
78.4	78.3	70.6	69.7	62.6	60.6	55.2	51.7	47.6	41.9
78.2	78.1	70.4	69.5	62.4	60.3	55.0	51.4	47.4	41.6
		70.2	69.3	62.2	60.1	54.8	51.2	47.2	41.3
78.0	77.8					54.6	50.9	47.0	41.1
77.8	77.6	70.0	69.0	62.0	59.9	54.4	50.7	46.8	40.8
77.6	77.4	69.8	68.8	61.8	59.6	54.2	50.4	46.6	40.5
77.4	77.2	69.6	68.6	61.6	59.4			46.4	40.3
77.2	77.0	69.4	68.4	61.4	59.1	54.0	50.2	46.2	40.0
77.0	76.8	69.2	68.1	61.2	58.9	53.8	49.9		
76.8	76.5	69.0	67.9	61.0	58.7	53.6	49.7	46.0	39.7
76.6	76.3	68.8	67.7	60.8	58.4	53.4	49.4	45.8	39.4
76.4	76.1	68.6	67.5	60.6	58.2	53.2	49.2	45.6	39.2
76.2	75.9	68.4	67.2	60.4	58.0	53.0	48.9	45.4	38.9
		68.2	67.0	60.2	57.7	52.8	48.6	45.2	38.6
76.0	75.7					52.6	48.4	45.0	38.3
75.8	75.5	68.0	66.8	60.0	57.5	52.4	48.1	44.8	38.1
75.6	75.2	67.8	66.6	59.8	57.2	52.2	47.9	44.6	37.8
75.4	75.0	67.6	66.3	59.6	57.0			44.4	37.5
75.2	74.8	67.4	66.1	59.4	56.8	52.0	47.6	44.2	37.2
75.0	74.6	67.2	65.9	59.2	56.5	51.8	47.4		
74.8	74.4	67.0	65.6	59.0	56.3	51.6	47.1	44.0	36.9
74.6	74.1	66.8	65.4	58.8	56.1	51.4	46.9	43.8	36.7
74.4	73.9	66.6	65.2	58.6	55.8	51.2	46.6	43.6	36.4
74.2	73.7	66.4	65.0	58.4	55.6	51.0	46.3	43.4	36.1
		66.2	64.7	58.2	55.3	50.8	46.1	43.2	35.8
74.0	73.5					50.6	45.8	43.0	35.5
73.8	73.3	66.0	64.5	58.0	55.1	50.4	45.6	42.8	35.3
73.6	73.0	65.8	64.3	57.8	54.8	50.2	45.3	42.6	35.0
73.4	72.8	65.6	64.0	57.6	54.6			42.4	34.7
73.2	72.6	65.4	63.8	57.4	54.4	50.0	45.0	42.2	34.4
73.0	72.4	65.2	63.6	57.2	54.1	49.8	44.8		
72.8	72.2	65.0	63.3	57.0	53.9	49.6	44.5	42.0	34.1
72.6	71.9	64.8	63.1	56.8	53.6	49.4	44.3	41.8	33.8
72.4	71.7	64.6	62.9	56.6	53.4	49.2	44.0	41.6	33.5
72.2	71.5	64.4	62.7			49.0	43.7	41.4	33.2
		64.2	62.4					41.2	32.9

$k = 80$

s	l	s	l	s	l	s	l	s	l
41.0	32.7	33.0	20.1	25.0	5.3	17.0	—13.8	9.0	— 42.7
40.8	32.4	32.8	19.8	24.8	4.8	16.8	—14.4	8.8	— 43.6
40.6	32.1	32.6	19.5	24.6	4.4	16.6	—15.0	8.6	— 44.6
40.4	31.8	32.4	19.1	24.4	4.0	16.4	—15.6	8.4	— 45.6
40.2	31.5	32.2	18.8	24.2	3.6	16.2	—16.1	8.2	— 46.7
40.0	31.2	32.0	18.4	24.0	3.2	16.0	—16.7	8.0	— 47.7
39.8	30.9	31.8	18.1	23.8	2.7	15.8	—17.3	7.8	— 48.8
39.6	30.6	31.6	17.7	23.6	2.3	15.6	—17.9	7.6	— 49.9
39.4	30.3	31.4	17.4	23.4	1.9	15.4	—18.5	7.4	— 51.0
39.2	30.0	31.2	17.0	23.2	1.4	15.2	—19.1	7.2	— 52.2
39.0	29.7	31.0	16.7	23.0	1.0	15.0	—19.7	7.0	— 53.4
38.8	29.4	30.8	16.3	22.8	0.6	14.8	—20.4	6.8	— 54.6
38.6	29.1	30.6	16.0	22.6	0.1	14.6	—21.0	6.6	— 55.9
38.4	28.8	30.4	15.6	22.4	— 0.3	14.4	—21.6	6.4	— 57.1
38.2	28.5	30.2	15.3	22.2	— 1.0	14.2	—22.3	6.2	— 58.5
38.0	28.2	30.0	14.9	22.0	— 1.2	14.0	—22.9	6.0	— 59.8
37.8	27.9	29.8	14.5	21.8	— 1.7	13.8	—23.6	5.8	— 61.2
37.6	27.5	29.6	14.2	21.6	— 2.2	13.6	—24.3	5.6	— 62.6
37.4	27.2	29.4	13.8	21.4	— 2.6	13.4	—25.0	5.4	— 64.1
37.2	26.9	29.2	13.4	21.2	— 3.1	13.2	—25.6	5.2	— 65.7
37.0	26.6	29.0	13.1	21.0	— 3.6	13.0	—26.3	5.0	— 67.3
36.8	26.3	28.8	12.7	20.8	— 4.0	12.8	—27.0	4.8	— 68.9
36.6	26.0	28.6	12.3	20.6	— 4.5	12.6	—27.8	4.6	— 70.6
36.4	25.7	28.4	12.0	20.4	— 5.0	12.4	—28.5	4.4	— 72.4
36.2	25.4	28.2	11.6	20.2	— 5.5	12.2	—29.2	4.2	— 74.3
36.0	25.0	28.0	11.2	20.0	— 6.0	12.0	—30.0	4.0	— 76.2
35.8	24.7	27.8	10.8	19.8	— 6.5	11.8	—30.7	3.8	— 78.3
35.6	24.4	27.6	10.4	19.6	— 7.0	11.6	—31.5	3.6	— 80.4
35.4	24.1	27.4	10.0	19.4	— 7.5	11.4	—32.3	3.4	— 82.6
35.2	23.8	27.2	9.7	19.2	— 8.0	11.2	—33.1	3.2	— 85.0
35.0	23.4	27.0	9.3	19.0	— 8.5	11.0	—33.9	3.0	— 87.5
34.8	23.1	26.8	8.9	18.8	— 9.0	10.8	—34.7	2.8	— 90.2
34.6	22.8	26.6	8.5	18.6	— 9.5	10.6	—35.5	2.6	— 93.0
34.4	22.5	26.4	8.1	18.4	—10.0	10.4	—36.4	2.4	— 96.1
34.2	22.1	26.2	7.7	18.2	—10.6	10.2	—37.2	2.2	— 99.4
34.0	21.8	26.0	7.3	18.0	—11.1	10.0	—38.1	2.0	—103.0
33.8	21.5	25.8	6.9	17.8	—11.6	9.8	—39.0	1.8	—107.0
33.6	21.1	25.6	6.5	17.6	—12.2	9.6	—39.9	1.6	—111.4
33.4	20.8	25.4	6.1	17.4	—12.7	9.4	—40.8	1.4	—116.4
33.2	20.5	25.2	5.7	17.2	—13.3	9.2	—41.7	1.2	—122.1
								1.0	—128.8

TABLE 5.2

D-values belonging to *S*

The Gaussian subtraction logarithm values [17] $D = S - \lg(10^S - 1)$ belonging to the blackening values S

The same data also give the values of $d \equiv D_{s/\gamma} = s - \lg(10^s - 1)$ belonging to the reduced blackenings s

In the table the hundredfolds of the true values of S and D, and the corresponding s- and d-values are given

S	D	S	D	S	D	S	D
200	0.4	80	7.5	50	16.5	35.0	25.7
190	0.5	79	7.7	49.5	16.7	34.5	26.1
180	0.7	78	7.9	49	17.0	34.0	26.5
170	0.9	77	8.1	48.5	17.2	33.5	27.0
160	1.1	76	8.3	48	17.5	33.0	27.4
150	1.4	75	8.5	47.5	17.7	32.5	27.8
140	1.8	74	8.7	47	18.0	32.0	28.3
130	2.2	73	8.9	46.5	18.2	31.5	28.8
120	2.8	72	9.2	46	18.5	31.0	29.2
115	3.2	71	9.4	45.5	18.8	30.5	29.7
110	3.6						
105	4.1	70	9.7	45	19.0	30.0	30.2
		69	9.9	44.5	19.3	29.8	30.4
100	4.6	68	10.2	44	19.6	29.6	30.6
99	4.7	67	10.4	43.5	19.9	29.4	30.8
98	4.8	66	10.7	43	20.2	29.2	31.0
97	4.9	65	11.0	42.5	20.5	29.0	31.2
96	5.0	64	11.3	42	20.8	28.8	31.4
95	5.2	63	11.6	41.5	21.1	28.6	31.7
94	5.3	62	11.9	41	21.4	28.4	31.9
93	5.4	61	12.2	40.5	21.7	28.2	32.1
92	5.6						
91	5.7	60	12.6	40.0	22.0	28.0	32.3
		59	12.9	39.5	22.4	27.8	32.5
90	5.8	58	13.3	39.0	22.7	27.6	32.8
89	6.0	57	13.6	38.5	23.1	27.4	33.0
88	6.1	56	14.0	38.0	23.4	27.2	33.2
87	6.3	55	14.4	37.5	23.8	27.0	33.4
86	6.5	54	14.8	37.0	24.2	26.8	33.7
85	6.6	53	15.2	36.5	24.5	26.6	33.9
84	6.8	52	15.6	36.0	24.9	26.4	34.2
83	7.0	51	16.1	35.5	25.3	26.2	34.4
82	7.1						
81	7.3						

S	*D*	*S*	*D*	*S*	*D*	*S*	*D*
26.0	34.6	19.4	44.3	12.8	59.3	6.0	88.9
25.8	34.9	19.2	44.7	12.6	59.9	5.8	90.3
25.6	35.1	19.0	45.1	12.4	60.5	5.6	91.7
25.4	35.4	18.8	45.4	12.2	61.1	5.4	93.2
25.2	35.6	18.6	45.8			5.2	94.8
25.0	35.9	18.4	46.2	12.0	61.7	5.0	96.4
24.8	36.1	18.2	46.6	11.8	62.4	4.8	98.0
24.6	36.4			11.6	63.0	4.6	99.8
24.4	36.7	18.0	46.9	11.4	63.7	4.4	101.6
24.2	36.9	17.8	47.3	11.2	64.3	4.2	103.6
		17.6	47.7	11.0	65.0		
24.0	37.2	17.4	48.1	10.8	65.7	4.0	105.6
23.8	37.5	17.2	48.5	10.6	66.4	3.8	107.7
23.6	37.8	17.0	49.0	10.4	67.2	3.6	109.9
23.4	38.0	16.8	49.4	10.2	67.9	3.4	112.3
23.2	38.3	16.6	49.8			3.2	114.9
23.0	38.6	16.4	50.2	10.0	68.7	3.0	117.6
22.8	38.9	16.2	50.7	9.8	69.5	2.8	120.5
22.6	39.2			9.6	70.3	2.6	123.6
22.4	39.5	16.0	51.1	9.4	71.1	2.4	126.9
22.2	39.8	15.8	51.6	9.2	71.9	2.2	130.6
		15.6	52.0	9.0	72.8		
22.0	40.1	15.4	52.5	8.8	73.7	2.0	134.7
21.8	40.4	15.2	53.0	8.6	74.6	1.8	139.1
21.6	40.7	15.0	53.5	8.4	75.5	1.6	144.2
21.4	41.0	14.8	53.9	8.2	76.4	1.4	149.9
21.2	41.3	14.6	54.4			1.2	156.5
21.0	41.6	14.4	54.9	8.0	77.4	1.0	164.3
20.8	42.0	14.2	55.5	7.8	78.4	0.8	173.9
20.6	42.3			7.6	79.4	0.6	186.3
20.4	42.6	14.0	56.0	7.4	80.5	0.4	203.7
20.2	43.0	13.8	56.5	7.2	81.6	0.2	233.8
		13.6	57.0	7.0	82.7	0.0	∞
20.0	43.3	13.4	57.6	6.8	83.9		
19.8	43.6	13.2	58.2	6.6	85.1		
19.6	44.0	13.0	58.7	6.4	86.3		
				6.2	87.6		

TABLE 5.3

T_P-values belonging to S

(For notations *cf.* Section 1.2, List of Symbols)

In the table index P is omitted, furthermore the hundredfolds of the true S- and T-values are given

S	T	S	T	S	T	S	T
0	100.0	36	43.7	70	20.0	106	8.70
1	97.8	37	42.7	71	19.5	107	8.51
2	95.5	38	41.7	72	19.1	108	8.31
3	93.3	39	40.7	73	18.6	109	8.12
4	91.2			74	18.2		
5	89.1	40	39.8	75	17.8	110	7.94
6	87.0	41	38.9	76	17.4	111	7.76
7	85.1	42	38.0	77	17.0	112	7.58
8	83.1	43	37.2	78	16.6	113	7.42
9	81.2	44	36.3	79	16.2	114	7.25
		45	35.5			115	7.08
10	79.4	46	34.7	80	15.8	116	6.92
11	77.6	47	33.9	81	15.5	117	6.76
12	75.8	48	33.1	82	15.1	118	6.61
13	74.2	49	32.4	83	14.8	119	6.46
14	72.5			84	14.5		
15	70.8	50	31.6	85	14.2	120	6.31
16	69.2	51	30.9	86	13.8	121	6.16
17	67.6	52	30.2	87	13.5	122	6.03
18	66.1	53	29.5	88	13.2	123	5.90
19	64.6	54	28.8	89	12.9	124	5.76
		55	28.2			125	5.62
20	63.1	56	27.5	90	12.6	126	5.50
21	61.6	57	26.9	91	12.3	127	5.37
22	60.3	58	26.3	92	12.0	128	5.24
23	59.0	59	25.7	93	11.7	129	5.13
24	57.6			94	11.5		
25	56.3	60	25.1	95	11.2	130	5.01
26	55.0	61	24.5	96	11.0	131	4.90
27	53.7	62	24.0	97	10.7	132	4.79
28	52.4	63	23.4	98	10.5	133	4.68
29	51.3	64	22.9	99	10.2	134	4.57
		65	22.4			135	4.47
30	50.1	66	21.9	100	10.00	136	4.37
31	49.0	67	21.4	101	9.78	137	4.37
32	47.9	68	20.9	102	9.55	138	4.17
33	46.8	69	20.4	103	9.33	139	4.07
34	45.7			104	9.12		
35	44.7			105	8.91		

S	*T*	*S*	*T*	*S*	*T*	*S*	*T*
140	3.98	160	2.51	180	1.58	200	1.00
141	3.89	161	2.45	181	1.55	205	0.89
142	3.80	162	2.40	182	1.51	210	0.79
143	3.72	163	2.34	183	1.48	215	0.71
144	3.63	164	2.29	184	1.45	220	0.63
145	3.55	165	2.24	185	1.42	225	0.56
146	3.47	166	2.19	186	1.38	230	0.50
147	3.39	167	2.14	187	1.35	235	0.45
148	3.31	168	2.09	188	1.32	240	0.40
149	3.24	169	2.04	189	1.29	245	0.36
150	3.16	170	2.00	190	1.26	250	0.32
151	3.09	171	1.95	191	1.23	260	0.25
152	3.02	172	1.91	192	1.20	270	0.20
153	2.95	173	1.86	193	1.17	280	0.16
154	2.83	174	1.82	194	1.15	290	0.13
155	2.82	175	1.78	195	1.12	300	0.10
156	2.75	176	1.74	196	1.10	310	0.08
157	2.69	177	1.70	197	1.07	320	0.06
158	2.63	178	1.66	198	1.05	330	0.05
159	2.57	179	1.62	199	1.02	340	0.04
						350	0.03
						370	0.02
						400	0.01
						∞	0.00